D0276684

The Premonitions Bureau

I

The music school was in an ordinary terraced house on one of the main roads leading out of London, to the north. The front was pebble-dash, like its neighbour; there were lace curtains and neat, cared-for roses growing under the bay windows. A curved archway of red bricks framed the front door, to the left of which hung a black sign, with gold lettering in confidently varying fonts:

Miss Lorna Middleton

Teacher of Pianoforte &

Ballet   Dancing

69 Carlton Terrace – New Cambridge Road

Hardly anyone called her Lorna. Her first name was Kathleen; she signed letters Kathy or Kay. She was, to almost everyone that knew her, Miss Middleton. She played the piano beautifully, with small hands. She had dark wavy hair, buck teeth and a pronounced New England accent, which combined with a degree of innate personal magnetism to make Miss Middleton an object of some fascination in post-war Edmonton. Pupils joined her classes from the ages of three or four. Many remembered her as a singular figure for the rest of their lives.

Miss Middleton never just walked into a room, or stood around. She moved. She posed. Her school followed a syllabus that, she claimed, had much in common with the tuition on offer at Trinity College, the Guildhall and the Royal Academy. But each of her dance classes began with her rolling up the carpet in the front room and shifting chairs out of the way, while six girls, and occasionally a boy, filed in and found a place to practise their port de bras while leaning on a bookcase. Miss Middleton played the piano with her back to her pupils, swivelling on a stool. The furniture around her was dark and somewhat distinguished. A leather sofa with brass studs sat under the window, a note of inherited wealth that was out of keeping with the cheap reproductions of cavaliers and shy eighteenth-century beauties hanging on the walls, and the paper notice, warning of missed classes and late payment, taped to the glass of a display cabinet. Out in the hall, the next class waited on the stairs, trying to stay out of the way of Miss Middleton's small, fierce mother, Annie, who had once been a great beauty and, it was rumoured, a courtesan in Paris.

Miss Middleton called her pupils the Merry Carltons. Several times a year, she would stage ambitious school performances, which caused her great anxiety. Annie would sew the costumes while Miss Middleton would rehearse pieces with as many as forty children, as well as an ensemble to be performed by the group, perhaps a musical comedy, which she regarded as her great love. During the preparation for these shows, the Merry Carltons would be reminded, more than once, that Miss Middleton had enjoyed a dancing career of her own. The front room at Number 69 was scattered with performance programmes with the dates carefully removed: a newspaper clipping from the time she danced on Boston Common, before a crowd of fifty thousand; a photograph of a young woman performing a grand jeté, by a 'Bruno of Hollywood'.

Nothing was ever spelled out. Miss Middleton's pupils only shared a sense of something grand that never quite happened and the understanding, which developed over time, that their teacher's ambitions exceeded their own. Miss Middleton often parted ways with her students when they became teenagers and started to take their lessons less seriously. In turn, her pupils noticed that they rarely saw Miss Middleton outside the front room of Number 69. They picked up gossip that her American accent might be affected or put on. She was not someone you saw grocery shopping in Edmonton Green. Although she was not old (how old, it was genuinely impossible to say), it was obvious that the great hopes of Miss Middleton lay in the past, that her true dreams had gone unrealised.

Late in her life, Miss Middleton typed out a list of her instructions for teaching music. The intended readership was unclear. Rule number five is addressed to pupils: 'Do not play by ear.' Number seven is a teaching tip: 'Octaves should be taught as soon as possible.' Number nine is blank. Many of the rules are not really rules but Miss Middleton's observations or personal entreaties.

12. Play as accurately and as well as possible bear in mind the teacher can get a headache and lose patience as well as the pupil.

22. Story of pupil who wore gloves while practising.

26. Do not keep repeating everything.

*

On a cold winter's day, when Miss Middleton was about seven years old, she came home from school for lunch and watched her mother frying eggs on the stove. 'After about two minutes, and without warning the egg lifted itself up. It rose up and up until it almost touched the ceiling,' Miss Middleton wrote in a self-published memoir, which appeared in 1989. She was excited by the sight and raced back to school to tell her friends. 'By the time I had re-told the story a thousand times the kids expected me to take off and fly into the clouds,' she wrote. But Annie was concerned. She consulted a fortune teller, who told her that an egg that flew out of the pan symbolised the death of someone close to you. A few weeks later, one of Annie's best friends, who had recently married, died and was buried in her wedding dress.

'I cannot say what I really felt or indeed what I feel now,' Miss Middleton wrote. She experienced premonitions, in one form or another, throughout her life. She compared

for Polly

First published in 2022
by Faber and Faber Ltd
Bloomsbury House,
74–77 Great Russell Street
London WC1B 3DA

Typeset by Typo•glyphix, Burton-on-Trent DE14 3HE
Printed and bound by CPI Group (UK) Ltd, Croydon, CR0 4YY

Quote on page 71 from *G* by John Berger, 1972, published by
Weidenfeld & Nicolson; new edition, 2012, published by Bloomsbury
Quote on page 173 from 'New York Mining Disaster 1941',
songwriters Barry Gibb and Robin Gibb, 1967

The right of Sam Knight to be identified as author of this work
has been asserted in accordance with Section 77 of the Copyright,
Designs and Patents Act 1988

A CIP record for this book
is available from the British Library

ISBN 978–0–571–35756–7

MIX
Paper from
responsible sources
FSC® C171272

6 8 10 9 7 5

The Premonitions Bureau

A TRUE STORY

Sam Knight

faber

the feeling to knowing the answer in a spelling test. Names and numbers would appear to her. 'I am drawn to these events by what appears to be a blaze of light,' she wrote. 'An electric light bulb.' When Miss Middleton was eleven, she felt an irresistible urge to contact her piano teacher, a young German man, who had recently been hospitalised for nerve trouble. After cajoling her parents to call him, she found out that he had poisoned himself in his apartment. 'It was probable that fate would have intervened and his moment of death was there,' she reasoned. 'But I could not rid myself of the thought that if I had managed to contact him he would have returned for supper and any problems could have been discussed.' Miss Middleton was an only child and she sensed a world that was particularly responsive and legible to her. 'Everything happened just as I knew it would,' she wrote to a cousin. Her mother asked her to stop saying what would happen next.

Miss Middleton considered her childhood to be the happiest time of her life. She liked to reminisce about the 'large house of twelve rooms' where she had lived, and how her father 'had been offered a position' in America. The truth was much more modest. Annie and Henry, her father, were English. Henry came from a prosperous family which owned a furniture-making business and thirty properties across Islington and Hackney, in north London. Annie was one of five children from Liverpool. They met in Paris, not long before the First World War, and sailed to America on a ship named the *Bohemia* in something of a scandal. (Annie

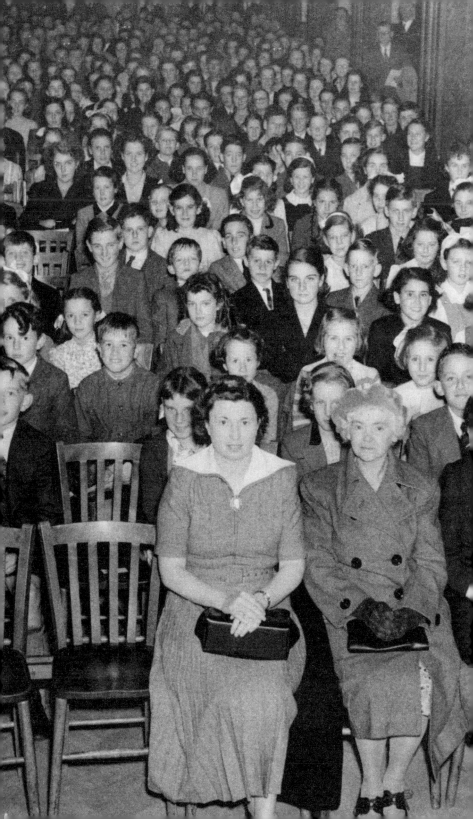

left behind an infant son in France.) In Boston, where Miss Middleton was born in 1914, Henry worked on the north docks, as a machinist at a canned goods store that was known for its devilled ham. The family lived in Dorchester, on the edge of the city. Guided by Annie, Miss Middleton took piano, dance and elocution lessons. She had a Russian ballet teacher and went to a progressive high school, where she learned dress design and how to fix cars and radios. She had a friend, Gloria Gilbert, who made it to Hollywood, where she became known as 'The Human Top', for her spins. But Henry's work dried up. In 1933, the family sailed back across the Atlantic, pursued by debts.

The return to England was humiliating. Carlton Terrace was a street of furriers, paper cutters and carpenters, a quiet, suburban realm quite apart from Paris, Hollywood and the rest of the Middleton family. At the age of fifty, Henry found work as a lathe operator. Less became possible. Miss Middleton auditioned at Sadler's Wells but could not afford the necessary tuition. When the Second World War began, she was working as a dance teacher at Prince's Dance Hall, a mile and a half away across north London, in Palmers Green. She took piano lessons by candlelight from an elderly organist named Mr E. A. Crusha, whose windows had been blown out during an air raid.

On a Saturday night in March 1941, Miss Middleton was preparing to go out for the first time since the start of the Blitz the previous autumn. There was a St Patrick's day celebration at Prince's Dance Hall. The place would be crowded with people that she knew. The air raid sirens had sounded and there was the rumble of bombs falling, but Miss Middleton was determined to go. She was just about to leave the house when a friend stopped by. They discussed whether it was safe to head out, and Miss Middleton decided that they would.

It was only after setting off that Miss Middleton experienced what she later described as 'a most strange sensation'. She took her friend's arm and they returned home, where they sat and played cards with Annie. While they were playing, at 8.45 p.m., a German bomber was hit by anti-aircraft fire and jettisoned its payload of high explosive over Palmers Green. Prince's was filled with dancers. A sixteen-year-old girl named Wyn was sitting with her friends, watching couples turn in front of her, when she felt a great rush of wind as the side of the building came off. 'You don't hear anything. That was when the bomb dropped,' she told the BBC in an interview. 'Everything went dark.' A sailor called out, telling people to stand against the walls. Wyn was pulled from the rubble. The casualties from the dance were laid out on the pavement outside. Only two people had been killed. Outside the hall, however, on Green Lanes, an electric trolley bus had been caught in the heart of the explosions. A fire-fighter, George Walton, arrived within moments and boarded the bus, which had been on its way to Southgate Town Hall. Forty-three passengers, quite dead, were sitting, standing and reading their newspapers, waiting for their stop.

*

It was not unusual, during the Blitz, to believe that your
life had been saved, or altered, by a premonition. The shattered streetscapes and possibility of death made the city
an uncanny place, in which it was not necessarily easy to
delineate what was real and what only existed in people's
minds. During almost nightly bombing raids, Londoners
sought sense, and solace, where they could find it. A fire
watcher, whose job it was to look out for falling bombs
and put out small fires, noticed that whenever he cleaned
his rubber boots, a bad night seemed to follow. So he left
them dirty.

In the spring of 1942, a survey by Mass Observation, a
social research organisation that was set up to record experiences of daily life in Britain, asked people about their
beliefs in the supernatural. Around a quarter of respondents
believed in some form of the occult, roughly the same proportion who thought there was an afterlife. Many challenged
the premise of the question, asking how it was possible to
distinguish between what was magical and what was simply
yet to be understood. 'I don't know where the "supernatural" begins and the "subconscious" ends,' a fifty-one-year-old
teacher from Barnet replied. While categorically spooky
things like ghosts, or ectoplasm, belonged in the realm of the
supernatural, there was much less agreement in mid-century
Britain about things like telepathy and apparently common
occurrences, such as premonitions, which hinted instead

9

at undiscovered reaches of physics and of the mind. One respondent to the Mass Observation survey wrote:

> I sometimes have very strong sensations that certain happenings will take place. Without rhyme or reason I *know*. At times the feeling has a logical background at others none at all. Until recently I didn't consider them or notice them seriously until after the event had occurred. Now I remark at the time and find that the expected results occur.

In the summer of 1944, the somewhat predictable terrors of the night-time raids of the Blitz were replaced by the haphazard arrival of flying bombs. The German V1, and later V2, rockets could strike at any time of the day or night. For many Londoners, frayed by five years of war, the flying bombs were more terrifying than anything they had experienced before. In order to confuse German spies and the rocket-aiming crews in northern Europe, misinformation about the times they came down was given to the newspapers. It was hard to make sense of what was happening. Citizens devised theories about which parts of the city were safe, which were not, whether the rockets could be aimed and if they fell in clusters. 'I wasn't frightened in the war until the rockets started coming over,' Wyn, the girl who survived the dance hall bombing, recalled in her interview with the BBC. A V2 destroyed an army uniform factory in Edmonton, not far from where she and Miss

Middleton lived. 'It was lucky it was the night,' Wyn said. 'It was absolutely flattened.' In 1946, Roland Clarke, an actuary at the Prudential Assurance Company, who had worked in military intelligence studying V1s during the war, published a one-page paper describing their distribution across London. He showed that for 144 square kilometres across the south of the city, where most of the rockets landed, the V1s struck almost perfectly randomly, fitting a mathematical formula named Poisson's Law, which had been used equally well in 1898 to calculate the number of Prussian soldiers accidentally kicked to death by horses.

*

By the mid-1960s, Miss Middleton had been teaching in the front room of Number 69 for almost a quarter of a century. When Henry and Annie were in their seventies, they inherited four houses in Holloway, a working-class neighbourhood in north London, and died not long afterwards. Miss Middleton kept cats, which multiplied. At one point, Les Bacciarelli, a Polish émigré who worked for the Post Office, moved in. Bacciarelli was an old flame of Miss Middleton's from the war. He became her lifelong companion. She described him as her lodger.

Premonitions continued to inform and change the direction of her life. After her mother died, Miss Middleton pursued an intuition which had first occurred to her as a child – that Annie's long-abandoned son lived in a pretty house by a river in France. In 1962, with the help of the US Embassy in Paris, Miss Middleton found her half-brother, Alexander, living in an old house in a small town on the banks of the river Sarthe, to the southwest of the city.

She never worked as a psychic or seemed unduly bothered by her sensations. 'I see no reason why this gift should be any more frightening than having a good head for mathematics,' Miss Middleton would say. She would bring out sketches of recent visions to show her music students and occasionally complain about all the information reaching her. 'She would say sometimes, "I just turn it off. I am too busy. I am too busy,"' Christine Williams, a former pupil, recalled. 'And she would wave her hand.'

*

Seeing patterns where they do not exist is known as apophenia. Finding meaning where there is none is a neat definition of madness. (A German neurologist, Klaus Conrad, came up with the label in 1958 when describing the origins of schizophrenia.) But making connections, in what we see, or hear, or dream, is also a definition of thought itself and finding a pattern that no one has ever found before, in computational physics, or in song (as long as someone else sees it, or feels it, too), is what we recognise as genius. 'They who dream by day are cognizant of many things which escape those who dream only by night,' Edgar

Allan Poe wrote in 1875. 'In their grey visions they obtain glimpses of eternity, and thrill, in waking, to find that they have been upon the verge of the great secret.' Just over a hundred years later, Barbara Brundage, a mental-health nurse in Minnesota, described how she recognised the start of one of her own psychotic episodes: 'I have a sense that everything is more vivid and important; the incoming stimuli are almost more than I can bear. There is a connection to everything that happens – no coincidences. I feel tremendously creative.'

*

On the night of 20 October 1966, when she was fifty-two, Miss Middleton decided to stay the night at one of her parents' inherited properties, on Crescent Road in Hornsey. She was restless and slept in a spare bedroom on the first floor. The next morning, at around 6 a.m., she had a powerful feeling of foreboding. 'I awoke choking and gasping and with the sense of the walls caving in,' she wrote soon afterward. Miss Middleton told Bacciarelli about the ominous feeling when he came home that morning, from a night shift. Bacciarelli found her depressed. At 8 a.m., Miss Middleton accepted a cup of tea, although she rarely drank anything in the morning.

A little more than an hour later, a group of labourers who were working on an enormous heap of coal waste in south Wales also paused to make a cup of tea. The team had a lightweight shack, with a coal fire, which they moved around on the hillside, depending on where they were working. It was a Friday morning – bright, windless, autumnal. The valley below was hidden under a layer of mist, which was pierced by the tall, square chimney of the Merthyr Vale Colliery. Since the First World War, spoil from the coal mine had been hauled up the side of Merthyr Mountain on a tramway. The waste, which included boiler ash, mine rubbish, discarded coal, slurry (small pieces of coal mixed with water) and tailings (much finer particles, left over from a chemical filtration process), came up the rails in sets of ten metal trams, which were pulled by a rope. When the trams reached the top of the slope and an engine hut, they slid gently down a separate track to the summit of the tip, where a team of men, known as slingers, attached each tram to a crane and a crane driver swung the tram over the waste heap and turned it upside down. Hour by hour, tram by tram, the heaps grew, dark and conical, high on the valley rims. When a tip became too large, or caused trouble on the hillside, the engineers at the colliery would find a site for a new one. Tip number seven, where the men were working that morning, had been started in Easter 1958, after a local farmer complained that the previous tip, number six, had begun to encroach on his fields. The spot was chosen by an engineer and the colliery manager, who walked up Merthyr Mountain one day without a map.

Twice in 1963, waste on tip number seven had slipped down the hillside. That November, a hole opened in the heap which was eighty yards wide. By the autumn of 1966, tip number seven rose 111 feet above the slope. It contained enough slurry, tailings and mine detritus to fill St Paul's Cathedral one and a half times over. Weeks of heavy rain had saturated the hills and the coal waste balanced on top. When the slingers and the crane driver arrived at the summit, just before seven thirty on the morning of 21 October, they noticed that the surface had sunk about ten feet in the night. The tram tracks that led to the edge of the tip had fallen into a hole. A slinger, Dai Jones, was sent down the mountain to report the movement. There was no working telephone on the tip because the line had been stolen. While Jones was gone, Gwyn Brown, the crane driver, moved the crane back. By the time Jones returned with the leader of the team, Leslie Davies, at around 9 a.m., the top of the heap had fallen another ten feet. The men did not like what they were seeing. Davies passed on the news that the colliery engineers would select a new tipping site next week. Tip number seven was finished. Davies suggested that they make a round of tea before the men began the work of moving the crane and the rails. The slingers and Davies headed for the shack.

Brown stayed by the crane and looked down the hillside. The valley was still carpeted by fog. There was no view of the close-packed terraces, churches and small shops of Aberfan. The village below was isolated but it was not rural, or even old. Before the mine there had been a single farmhouse and fields dotted with sheep. The river Taff had glittered on its way to the sea. In the nineteenth century, men came from England, Ireland and Italy to dig coal in Aberfan. They brought their families and fashioned a place. The colliery had seen them through thick and thin. In 1934, the village jazz band won a national competition, held at the Crystal Palace in London.

As Brown glanced down, the heap rose up. It did not make sense. 'It started to rise slowly at first,' the crane driver later said. 'I thought I was seeing things. Then it rose up after pretty fast, at a tremendous speed.' Far below, at the base of the tip, thousands of tons of waste had liquefied and suddenly fallen. A dark, glistening wave burst out of the hillside and poured down, carrying the rest of the heap with it. 'It sort of came up out of the depression and turned itself into a wave – that is the only way I can describe it,' Brown said. 'Down towards the mountain . . . towards Aberfan village . . . into the mist.' Brown called out. The rest of the team stumbled from the shack, saw what was happening and ran down the slope, shouting warnings in the air. Their way was blocked by falling trees, trams, muck and slurry. The noise was tremendous. The men had all seen tip slides of a few hundred yards or so. But this was an avalanche. People in Aberfan later compared the roar of the tip to a low-flying jet aircraft or thunder as it swept down.

Sheep, hedges, cattle, a farmhouse with three people inside were smothered. The westernmost street in the village, which lay against the mountainside, was Moy Road. It was where Aberfan's two schools were situated: Pantglas Junior School and Pantglas County Secondary School. Classes at

16

the junior school began at nine o'clock. The senior school got going at nine thirty. The wave reached them at nine fifteen, burying the primary school, which was full of children answering the register, checking the rain gauge, spelling the word p-a-r-a-b-l-e, paying their dinner money, delivering school sports reports, preparing to draw. Colliery trams and boulders crashed through the walls. The rear of the school was crushed beneath a dark heap thirty feet high. The gable ends of the roof poked through the waste. The senior school was only partially hit. Howard Rees, a fourteen-year-old boy on his way to class, saw the wave crest an old railway embankment above the village, 'moving fast, as fast as a car goes into town,' and crush three friends who were sitting on a wall. Eight houses behind them went too. George Williams, a hairdresser, saw doors and windows crashing inwards on Moy Road. Bricks flew. He was pinned under a piece of corrugated iron. When the roar stopped, Williams likened the sound to the moment after you switch a radio off. 'In that silence you couldn't hear a bird or a child,' he said. The first emergency call, from the Mackintosh Hotel, a pub further down Moy Road, was timed at 9.25 a.m. Miners, their faces blackened, lamps on their helmets, came up from the coal seams under the valley and were on the scene in twenty minutes. Water from severed pipes rushed through the streets, up to the rescuers' knees. One hundred and forty-four people were killed by the tip slide in Aberfan, 116 of them children, mostly between the ages of seven and ten.

The BBC broadcast a newsflash at 10.30 a.m. On the lunchtime news, which is how the prime minister, Harold Wilson, heard of the disaster, the death toll was given as twenty-six. By that time, Aberfan, which is tucked off the main road between Merthyr Tydfil and Cardiff, was becoming clogged with press vehicles, ambulances, mobile canteens and earth-moving machinery. Nearby mines sent all manner of tractor shovels, bulldozers, excavators and trucks to clear the debris, but the tight corners of the school-yard and the possibility of survivors in the slurry meant that the search was done almost entirely by hand. Every time a rescuer thought they had found something, a whistle blew and the place fell silent. No one was pulled alive from the wreckage after eleven o'clock in the morning. A dead girl was found holding an apple, a boy clutched at four pence. Children were found with birth certificates folded in their pockets. Some bodies were terribly disfigured. There was frenzy in the desire to help, to undo what had just occurred. People wanted to be useful even though it was not possible. At Merthyr hospital, people queued up to give blood, despite the lack of need. The switchboard at the colliery was jammed by incoming calls offering help in one form or another, making it impossible to find an out-side line. Between one and two thousand people hurried to join the dig on Moy Road. Men cut themselves and bled in the muck. People stood on the waste heap, watching the effort, and caused it to crumble further, delaying the recovery. A bulldozer driver fell asleep at the controls. On the hillside above the village, miners and engineers worked to stabilise the rest of tip number seven with sandbags, which they filled with the slurry all around them. Near the school, around a hundred off-duty ambulance drivers hung around, refusing to go home. During the afternoon, higher-wattage bulbs were fitted in Aberfan's street lamps, floodlights were erected and the digging went on. Darkness fell and it grew cold. The prime minister came and went. Lord Snowdon, the Queen's brother-in-law, arrived at around 3 a.m. with a small suitcase and a shovel and was taken to the Bethania Chapel, the largest in the village, where the bodies were laid out on dark wooden pews marked by chalked letters – 'M' for male, 'F' for female, 'J' for juvenile – and a group of police detectives worked. Outside, a line of around fifty parents, mostly fathers, waited for hours to identify their dead.

The following morning, which was a Saturday, was overcast. The clouds spelled rain. Most people in the village had slept for only an hour or two. 'There was a greyness everywhere,' the *Merthyr Express* reported. 'Faces from the tiredness and anguish, houses and roads from the oozing slurry of the tips.' There was more order and less hope, but the manic energy of the previous day persisted. A hundred lorries queued through the village to take away the mess. In twenty-four hours, the world had learned the name of Aberfan and its meaning: a place whose children had been buried alive by coal waste piled up by their fathers. The Red Cross handed out ten thousand cigarettes. The RSPCA despatched

a mobile unit and five animal inspectors, who went house to house, checking if pets had been disturbed by the chaos and needed looking after. (None did.) If anything was needed at the disaster site, it arrived rapidly and in hysterical quantities. When a request went out for gloves, six thousand pairs turned up. The police asked for a piece of digging equipment known as a '955 shovel'; they received 955 shovels instead. Lorries that no one had sent for pulled into the colliery yard bearing tinned meat, shirts and tons of fruit. Chewing gum, soap, soup and bottles of brandy piled up wherever there was space.

*

There was a roadblock to control access to the disaster, but more or less anyone in a uniform or an official-looking car could find a way through. During the morning of 22 October, a dark green Ford Zephyr nosed its way into the village. At the wheel was John Barker, a forty-two-year-old psychiatrist with a keen interest in unusual mental conditions. Barker was tall and broad and dressed in a suit and tie. In his thirties, he had been very overweight. Since then, he had taken up exercise and dieted on rusks – hard, twice-cooked pieces of bread – with the result that his clothes now hung loose and he looked older than his years. He had bags under his eyes, fleshy lips and dark hair, which he wore combed forward. Barker was a senior consultant at Shelton Hospital, a mental institution outside Shrewsbury,

21

a hundred miles east of the Merthyr valley, on the other side of the border with England. At the time, he was working on a book about whether it was possible to be frightened to death.

In the early news reports from Aberfan, Barker had heard that a boy had escaped from the school unharmed but later died of shock. The psychiatrist had come to investigate, but realised he had arrived too soon. When Barker reached the village, victims were still being dug out. 'I soon realised it would have been quite inopportune to make any enquiries about this child,' he wrote afterwards. He was appalled by what he saw. Barker was married and had three young children of his own. 'The experience sickened me,' he wrote. The devastation reminded Barker of the Blitz, when he had been a teenager, growing up in south London, but the loss of life in Aberfan was worse for being so concentrated and the dead so young. 'Parents who had lost their children were standing in the street, looking stunned and hopeless and many were still weeping. There was hardly anybody I encountered who had not lost someone.'

Voyeurs and outsiders who came to Aberfan without good reason were easy to identify. Policemen who stood around drinking tea were shouted at. Someone threw a tobacco tin at a photographer and broke his camera flash. During the course of the day, a steady drizzle came down, soaking the hundreds of rescuers, muddying the streets, which were already inches deep in muck, and raising fears that the tip could suddenly fall again, causing another calamity. The village was dreadfully tense. First aid stations withdrew from the foot of the hill. A klaxon was readied to sound the alarm.

But Barker did not get back in his car and drive away. He had long been interested in subjects that struck others as macabre or inexplicable. He was, in every outward sense, an orthodox psychiatrist. He had studied at Cambridge University and at St George's Medical School in London. But he also chafed at the limits of his field. Barker believed that there was a 'new dimension' to psychiatry, waiting to be incorporated into mainstream science, if doctors could be persuaded to study problems and conditions that were mostly regarded as fringe or psychic. He was a member of Britain's Society for Psychical Research, which was founded in 1882 to investigate the paranormal, and for some years had been interested in the problem of precognition, when people seemed to know what was going to happen before it actually did. Barker was a modern doctor; he pursued his more esoteric inquiries with what he described as 'a conscious rationalism'. But he also understood that there was an involuntary aspect to his research and that he was driven by influences that were profound and impulsive. At crucial moments in his life, when Barker was faced with a boundary or a warning, he pressed on.

In Aberfan, Barker sensed that he was on the scene of something momentous, though he wasn't sure what. Talking to witnesses, he was struck by 'several strange and pathetic incidents' connected to the disaster. A school bus, carrying children from Merthyr Vale, had been delayed by the fog and

23

reached Moy Road after the tip fell. Their lateness saved their lives. A boy had overslept, apparently for the first time in his life, and was sent hurrying to school by his mother, in tears; he was crushed. Inane, unthinking decisions in the moments before the waste came down – a cup of tea before starting work, looking the wrong way, resting on a wall – spared lives and ended others.

Barker was interested in the nature of those decisions and what prompted them. Did people have rational fears or unconscious knowledge? The dark, unnatural tips above Aberfan had long played on local people's minds. Bereaved families spoke of dreams and portents. Weeks after the accident, the mother of an eight-year-old boy named Paul Davies, who died in Pantglas School, found a drawing of massed figures digging in the hillside under the words 'the end', which he had made the night before the slide.

Barker also heard the story of Eryl Mai Jones, a ten-year-old girl, 'not given to imagination', who had told her mother, Megan, two weeks before the collapse that she was not afraid of death. 'Why do you talk of dying, and you so young?' her mother replied. 'Do you want a lollipop?'

Then, according to a statement written by Glannant Jones, a local minister, signed by Eryl Mai's parents and later published by Barker:

The day before the disaster she said to her mother, 'Mummy, let me tell you about my dream last night.' Her mother answered gently, 'Darling, I've no time. Tell me

25

again later.' The child replied, 'No Mummy, you must listen. I dreamt I went to school and there was no school there. Something black had come down all over it!'

The next morning, Eryl Mai was buried in the school.

The psychiatrist had an eye for an experiment. In the days after his visit to Aberfan, Barker came up with an idea for an unusual study. Given the singular nature of the disaster and its total penetration of the national consciousness, he decided to gather as many premonitions as possible of the event and to investigate the people who had them.

Barker wrote to Peter Fairley, the science editor of London's *Evening Standard*, and asked him to publicise the idea. The men had met the previous year, when Fairley had written a double-page feature about Barker's work on fear and death. They seemed quite different. Barker could be brittle and somewhat acerbic. Fairley, six years younger, was charming, tubby and competitive. As a junior reporter, he and a colleague had drunk a bottle of Mâcon-Villages and tossed a coin in the Two Brewers, the pub opposite the *Evening Standard*'s headquarters in Shoe Lane, to decide which beats they should pursue. Fairley landed science and had spent the following years writing urgent, gung-ho stories about atomic energy, deep-sea diving experiments and the space race. He went on television and toured the world. For a time, he lived on a houseboat in Chelsea, in west London, and chased stories on a fold-up motorcycle. There was an omnivorous, relentless quality to Fairley.

When he died, in 1998, his widow and four children learned that he had had a second family for twenty years. One of his children speculated that Fairley always sought to stay on top of everything, to chase down every lead, so that he couldn't be found out in his own life.

Fairley was hardly unconventional. He had been the head boy at Sutton Valence, a public school in Kent, and a captain in the army. But he had a proselytiser's faith in science; that one day it would answer all the questions we have ever had. He was open to theories about telepathy, extraterrestrial life and the mystery of seeming coincidence. He owed a significant part of his career to a hunch. In April 1961, on the basis of little more than a warning to ships in the Pacific and a feeling that something was up, Fairley had predicted that the USSR was about to launch its first manned space flight. His story ran on the front page of the *Evening Standard*: 'First Spaceman – Trip Imminent'. Fairley was thirty years old. He drove home to Bromley, passing black banner headlines outside the newspaper kiosks, feeling sick with apprehension.

Yuri Gagarin flew into space two days later. Fairley fielded calls from his editor asking how on earth he had known. His pay was almost doubled. The space race would, in time, become the story of his life. At the end of the decade, Fairley would go on to present the moon landings on British television. At the time of the Aberfan disaster, however, he was recovering from a spell of blindness brought on by a combination of diabetes and overwork. Fairley's eyes had started to play up the previous December, when he was haring back and forth between Cape Canaveral in Florida and NASA's mission control in Houston, Texas, while covering Gemini 7, an attempt to bring two spaceships alongside each other in orbit. He returned to England exhausted and woke up on Christmas Day unable to see. Fairley was blind for three months. The rest forced him to stop and to examine certain questions in his life. When he was in despair, Fairley crept downstairs and listened to recordings of Winston Churchill's wartime speeches over and over again.

Thinking that his condition might be permanent, Fairley had an idea to record some descriptions of the American and Soviet space programmes for other blind people. One day in the spring of 1966, he was doing the drying up after lunch. He felt suddenly despondent and useless. Then the telephone rang. It was a radio producer named Ron Hall, who had a question: could Fairley record a long interview about the space race for blind patients in the National Health Service? Hall came to Fairley's home with his guide dog the following week to record the programme. 'You can call it coincidence,' Fairley recalled later. 'But once a few of these things happen you start to wonder.'

On 28 October, Fairley carried Barker's appeal in his 'World of Science' column, which ran on a Friday. 'Did anyone have a genuine premonition before the coal tip fell on Aberfan? That is what a senior British psychiatrist would like to know,' Fairley wrote. The article described the kinds of vision that Barker was interested in: 'a vivid dream', 'a

vivid waking impression', 'telepathy at the time of the disaster (affecting someone miles away)' and 'clairvoyance'. Fairley's column implied that the study was somehow unauthorised: Barker had asked to remain anonymous. The article reminded readers of apparent foreknowledge of the sinking of the *Titanic* in 1912 and the R101 airship disaster in 1930, and ran next to an artist's impression of a lunar taxi that NASA was developing to carry astronauts around on the moon.

The *Evening Standard* had a circulation of almost six hundred thousand. Miss Middleton liked to look through it in bed in the afternoon. She posted an account of her premonition on 1 November.

*

29

The letters ended up in Barker's office on the first floor of Shelton Hospital, a long red-brick building two miles west of Shrewsbury which was obscured from the road by a screen of tall pine trees. The place was a Victorian asylum, in the gothic style. The main entrance had a bell tower, a pretty oriel window, and steeply pitched dormer windows in the roof. The hospital windows, which had small lead panes, opened four inches wide. The architect was George Gilbert Scott, designer of the St Pancras Hotel in London. In 1843, he and his partner, William Moffatt, had been asked to build a hospital for sixty inmates on a hill by Shropshire's Society for Improving the Condition of the Insane.

Within forty years, Shelton housed more than eight hundred people. On maps, the much extended hospital resembled a stick insect, with unfolding symmetrical wings and corridors for its male and female populations. Shelton's buildings were set in fifteen acres of grounds behind high brick walls. For more than a century the hospital was a closed stopping place, to which judges and doctors despatched the senile and the strange from a large part of the rural Midlands and Welsh borders. Around a quarter of the patients came from Shrewsbury, but the rest were from farming communities and small market towns. Some spoke only Welsh, if they spoke at all. There were some Poles left over from the war.

The institution was built to last forever. Shelton had its own steam laundry, barbers, upholstery workshop and brewery, which made one per cent proof hospital ale. 'Everything that needed to be done could be done,' Harry

Sheehan, a former assistant chief male nurse, recalled. Patients assembled screwdrivers and took apart old telephones in the industrial therapy workshops. Beyond the chapel lay the kitchen gardens and a piggery. There were orchards for plums, pears and apples and a small farm that was worked mostly by hand. Inmates drove the hospital's pigs, sheep and Ayrshire cattle down the hill to Shrewsbury on market day. Up until the 1930s, nurses were brought to work by horse and cart. The hospital cricket pitch was acknowledged as one of the finest in the county. People did not get better.

The ingrained purpose of Shelton, and dozens of county asylums like it, was not to cure people who were mentally ill but to sequester them from the world. By the middle of the twentieth century, when Shelton's population briefly topped a thousand, around two thirds of the patients were 'chronics' – long-stayers, who were not expected to leave. Some patients, particularly older women who were poor, simply materialised at the hospital without paperwork or a diagnosis. They were pushed in by one county authority or another, and a bed was found. Around a third never received a visitor. In the sixties, there was a deaf and dumb man at Shelton who had been there for forty-five years. There were roughly fifty people with learning disabilities (around six per cent of the patient population) who were not mentally ill but lived permanently with psychotic men and women because they had been abandoned. Alcoholics from across Shropshire were sometimes thrown into Shelton for a few weeks to dry out. Many patients received no treatment of any kind. Around half of the long-stay male patients did nothing all day. They played cards and repeated their fixations over and over. The wards were populated by former vicars, auctioneers and businessmen. There was at least one Jesus, who would perform miracles if only the nurses would let him out. Almost everyone wore clothes that were stained beyond belief. They were, in almost every sense, marooned. Occasionally a nurse would give a patient the wrong prescription, an antidepressant instead of a sedative, and after years of silence, the person would begin to speak. Patients who had been locked up for decades were sometimes taken for a walk and were astonished to see the traffic passing on the busy road to Welshpool. One spring day in the mid-sixties, a husband came to Shelton to collect the body of his wife, who had been held there for twenty-one years. But there had been a mix-up. She was not dead. He went home again.

Doctors who worked at Shelton called it 'a remote bin' and 'a dumping ground'. In Shrewsbury, the hospital was known as 'the Mental'. In truth, the place was no better and no worse than many county asylums at that time. Shelton was in the throes of something like reform. In 1964, the locks on all but two of the wards were removed. The male wards were named after celebrated men from the county, such as Charles Darwin and A. E. Housman, and the female wards were named after trees because no one could find a list of famous Shropshire ladies. Railings and walls within the grounds were replaced with trees and bushes. There was a psychologist and occupational therapy; jazz bands came to perform.

But the hospital remained powerfully constrained by what it was. There were staff who were as institutionalised as the patients. The hospital superintendent, John Littlejohn, was a former colonial administrator, who was frequently ill and struggled to make decisions. A lot of energy was spent simply getting from one day to the next. The hospital consumed 865 pints of milk per day. The grounds were infested with feral cats, which were a source of ringworm. There was a single post box, by the front door, for patients, staff and the hospital itself, which was often filled to overflowing, meaning that letters went missing. Nurses smoked constantly, in part to block out Shelton's all-pervading smell: of a house, locked up for years, in which stray animals had occasionally come to piss. The kitchen had a butcher with an attitude problem and the laundry sometimes went wrong, meaning that patients were forced to wear socks shrunk to half their normal size. The cutlery had to be counted out and counted in again at every meal. People got electrocuted in the industrial therapy unit. Patients slipped out to the pub, got drunk and attacked a nurse who happened to be passing. The roof of the Dutch barn on the farm was leaking. The road to Mytton Villa, a halfway house for recovering male patients, was impassable in the rain. Neighbours complained that Shelton's overgrown trees blocked the sun from reaching their houses. When a patient absconded from the Mental, an alarm

would be raised and volunteers from across Shrewsbury would help search the surrounding fields and ditches.

'I didn't have any sense of anything being hidden or that there was any cruelty,' a porter who worked at Shelton during this period recalled. 'But it was a very, very scary place.' Before he started at the hospital, the porter had worked in a chicken factory. The shifts at Shelton were better paid. In the mornings, he delivered drugs to the wards in wicker baskets. 'People would be sitting in chairs, rocking backwards and forwards, and the nurses wandering around cleaning up after incontinent patients,' he said. Often there was very little to do. During spare hours, the porter would go down to the basement and read handwritten patient records, Shropshire manias and melancholias from the nineteenth century, at a small wooden desk. One of his jobs was to remove bodies. The hospital had a mortuary opposite the tailor's shop, with three marble slabs protruding from the red-brick walls. Around a dozen people died every month at Shelton, mostly from old age. But every few weeks, someone would throw themselves out of a bathroom window, or hang themselves from a tree by the cricket pitch, or escape and throw themselves in front of a car. The primary means of discharge was death.

*

Barker started work at Shelton in the summer of 1963. He joined a team of four consultants who were responsible for around two hundred patients each, as well as a geographic share of Shrewsbury and the surrounding countryside. Barker's patch was to the north and east. He saw outpatients in the towns of Whitchurch, Market Drayton and Telford. He was on the side of a revolution that was taking place in mental health. Britain's vast, crowded hospitals inspired inertia but in the sixties, political reform, pharmaceutical possibilities and the intellectual atmosphere of psychiatry made it a lively and contested field. Barker quickly latched on to David Enoch, another young consultant at Shelton, who was Welsh, and had long hair and a confident, beguiling manner. Enoch's father had been a miner; Enoch himself had served as an army officer in India during the country's partition before training as a psychiatrist. He had arrived a year earlier from Runwell Hospital, a progressive mental hospital in Essex, and had been taken aback by the squalor and the apathy of Shelton.

'The old chaps told me, "You can do anything, David. But don't ruffle us,"' Enoch said. 'The mind-set was, people came in through the walls and never went out again.' Enoch opened windows and rationalised his patients from across the hospital into a small number of wards, so it was actually possible to see them all. He encouraged people out of their beds. He asked for lockers, so patients could store their possessions, and spare clothes for the men and women who worked in messy jobs. Enoch inherited sixty-four outpatients in Oswestry. Over time he realised that many of them were quite well and attended their appointments out of politeness, for fear of disappointing him.

Barker's last job had been at Herrison Hospital in Dorset, which had successfully integrated hundreds of long-stay and short-stay patients. 'He and I were kindred spirits,' Enoch recalled. Together, the two psychiatrists worked to improve conditions at Shelton. They phased out 'straight' electroconvulsive therapy, in which ECT was administered without drugs. Enoch organised adult education sessions which proved as popular with the staff as with the patients. 'As a result of the two of us there, heaps of things happened,' Enoch said. 'We stimulated each other. The old ones went on doing their own thing.' In 1965, the pair co-wrote a paper for *The Lancet* which showed that many hospitals were misusing their powers under the country's mental health legislation; they were summoned to the Ministry of Health in London to discuss their findings. They made a day trip on the train, feeling important.

The two men also shared a fascination with what Enoch called 'psychiatric orchids' – the most unusual mental illnesses. In the late fifties, Barker had written his doctoral thesis about Munchausen's syndrome, whose sufferers compulsively feign disease or harm themselves in order to be admitted to hospital, often undergoing needless surgery. Enoch invited Barker to contribute a chapter on Munchausen's to his own larger study of rare conditions, *Some Uncommon Psychiatric Syndromes*, which was first published in 1967 and is now a classic textbook. In the original edition, the case studies of erotomania (obsessional love), Othello syndrome (a delusion of your lover's infidelity) and

Couvade syndrome (in which a man seems to experience pregnancy at the same time as his partner) have a fable-like, poetic quality. There is a doleful wonder at what the mind is capable of. Barker's chapter opened with a fateful quotation from Karl Menninger, an American psychiatrist and co-founder of the Menninger Foundation in Topeka, Kansas: 'It is true, nevertheless, that in the end each man kills himself in his own selected way, fast or slow, soon or late.' Enoch used Dostoevsky to introduce his study of the Capgras delusion, 'a rare, colourful syndrome in which the patient believes that a person, usually closely related to him, has been replaced by an exact double.' Enoch had treated a man at Runwell who believed his wife was a replica.

The young consultants at Shelton were comfortable with attention. Enoch, in particular, was good-looking and admired by junior members of the hospital staff. He required two telephones on his desk and was a regular guest on *Let's Face Facts*, a general knowledge show on regional television, which made him a minor celebrity in western England and Wales. The porter who worked at Shelton went on to medical school and later became a successful surgeon. 'If anything made me want to be a doctor, it was how much I sort of revered someone like David Enoch,' he said. Barker was hungry for publicity, too. He was a pro-lific correspondent in the letters pages of *The Lancet* and the *British Medical Journal*. At Herrison Hospital, he had arranged a two-week exchange for patients with Fulbourn Mental Hospital in Cambridge and had it covered by the

Sunday People. During 1965, Barker weighed all the brochures and junk mail that he received from drug companies (it came to twelve kilograms) and invited a photographer from the *Birmingham Post* to take a picture of him deluged at his desk. It made the front page.

There were times when Enoch considered Barker avaricious. 'We would talk about things in the coffee room and the next minute he would be writing about them,' Enoch recalled. But these were moments of irritation in a larger project to bring modern, experimental psychiatry to a stagnant place like Shelton. Enoch also acknowledged that Barker had research interests which were entirely his own. Most of Barker's published work was about aversion therapy, a technique that involved the use of electric shocks and nausea-inducing drugs to treat addictions and other unwanted behaviours.

Earlier in his career, Barker had treated a man who dressed up in his wife's clothes and was afraid of being prosecuted as a transvestite. Two engineers at Banfield Hospital in Surrey, where Barker worked, devised an electrified rubber mat. Behind a screen next to the admissions ward, doctors played a loud buzzer and shocked the man's feet and ankles while he put on the female clothes and took them off again. 'During the first four days of treatment it was obvious that the patient found the procedure unpleasant, arduous and stressful,' Barker and his colleagues reported in the journal *Behaviour Research and Therapy*. But it seemed to work. Barker experimented and he hoped.

40

He pushed for grand conclusions. He believed that one day aversion therapy could be used to change behaviour among everyone from speeding drivers to people who struggled with their weight.

At Shelton, Barker had an electrified fruit machine installed outside his office, on the first floor, which gave gambling addicts a seventy-volt shock instead of money when they won. With another doctor, Mabel Miller, he also filmed them as they went to the betting shop. The psychiatrists then played the footage back while the addicts received electric shocks with an apparatus originally designed for testing monkeys. Barker played footage of their families, too, describing their anger and distress.

'A lot of his thinking was a bit futuristic,' Harry Sheehan, the nurse, said. To many people at Shelton, Barker came across as authoritative and direct. 'He didn't suffer fools gladly,' Sheehan said. 'If he asked for something and you said you would get it, then he expected it there.' Like Enoch, Barker took advantage of the sluggish administration of the hospital to pursue his own agenda. 'If he had started on something, some therapy, some course of treatment, he would carry it through,' Sheehan continued. 'Nobody would say, "You are not doing that, you can't do that." He would get on with it.'

But Barker's competence masked a frailty. His appointment at Shelton had been rushed, and a demotion. At Herrison Hospital, Barker had been the medical superintendent. He had been given charge of the hospital at the age of thirty-eight. But he only lasted a few months in the role. One colleague remembered him only as a stout doctor, interested in his own career, who made little effort to befriend other members of the medical staff. There were rumours of a nervous breakdown and possible behavioural indiscretions. On occasions during his life, Barker suffered what he described as 'emotional stress almost to the point of "crack up"'. He had arrived at Shelton in a brittle state. His weight fluctuated. His eyes bulged. He didn't look altogether well. 'He came to Shrewsbury under a cloud,' Enoch said. 'When I first saw him, I thought he had gone to seed . . . People said, "Why did he come to Shelton?" For shelter, really.' Enoch was too grateful to have a like-minded colleague to ever ask Barker much about his past. 'He was very able,' he said. 'I didn't want to know.'

*

Barker received seventy-six replies to his Aberfan appeal. Two nights before the disaster, a sixty-three-year-old man named J. Arthur Taylor, from Stacksteads, a village on the edge of the Lancashire moors, dreamed that he was in Pontypridd, in south Wales. He had not been in the town for many years and he was trying to buy a book. He faced a large machine with buttons. 'Now I have never seen a computer. This may have been one; I just don't know,' Taylor wrote. 'Then, all of a sudden, while I was standing by this big machine, I looked up and saw ABERFAN written as

42

if suspended in white lettering against a black background. This seemed to last some minutes. Then I turned and looked the other way and I saw through a window rows of houses and everything seemed derelict and desolate.' Taylor did not recognise the word, even though he had driven past the village countless times, until he heard it on the radio on the day of the disaster.

In Plymouth, the evening before the coal slide, Constance Milder had a vision at a spiritualist meeting. Milder, who was forty-seven, told six witnesses that she saw an old schoolhouse, a Welsh miner, and 'an avalanche of coal' rushing down a mountain. 'At the bottom of this mountain of hurtling coal was a little boy with a long fringe looking absolutely terrified to death. Then for quite a while I "saw" rescue operations taking place. I had an impression that the little boy was left behind and saved. He looked so grief-stricken.' Milder recognised the boy later on the evening news.

A man in Kent was convinced for days before the Aberfan accident that there would be a national disaster on the Friday. 'It came to me as strongly as might come the thought that you have forgotten that it was your wife's birthday tomorrow,' wrote R. J. Wallington, of Rochester. When he arrived at work on 21 October, he told his secretary: 'Today's the day.'

In Hillingdon, on London's western limit, Grace Richardson, a thirty-year-old film technician, was bothered all week by an intermittent smell, earthy and decaying, which she recognised as the smell of death. About an hour before the disaster, she asked a colleague named George Jordan, who worked next to her, whether he could smell anything unusual. He said no. About fifteen minutes after the school was buried, Richardson jumped up from her chair, over-whelmed, and said that something terrible had happened. 'Her face was highly inflamed and she was breathing very heavily,' Jordan wrote in an accompanying note to Barker. 'Neither of us or anyone else in the machine room had mentioned or heard of any disaster.'

Barker wrote back to the percipients, as he called them, asking for details and witnesses. Of the sixty plau-sible premonitions, there was evidence that twenty-two were described before tip number seven began to move. The material convinced Barker that precognition was not unusual among the general population – he speculated that it might be as common as left-handedness. Among the col-lection, he considered the vision of Eryl Mai Jones, the girl who died in Aberfan, 'an example of pure precognition' and the detailed description of Constance Milder, at the spiritualist meeting in Plymouth, equally astonishing.

As a doctor, Barker was particularly drawn to seven correspondents whose premonitions were accompanied by physical as well as mental symptoms. Along with Richard-son, the film technician, these included Miss Middleton, who woke up choking, and a woman who worked in the research department of the Bank of London and South America, who experienced 'very strong positive "waves"' that destroyed her concentration every two hours in the build-up to the disaster. Most of the men and women in this group described themselves as habitual seers, whose premonitions had been borne out over the years.

One of the seven was Alan Hencher, who worked on the continental telephone exchange for the Post Office. Hencher's letter was dated 29 October, the day after Fairley's article appeared, but unlike many of the other correspondents, who came across as either uncertain or over-eager to be believed, Hencher sounded almost indifferent. 'I accept that I am able to foretell certain events but I have no idea how or why,' he wrote.

Twenty-four hours before the Aberfan disaster, Hencher was working overtime at the GPO's international switchboard in Faraday House, not far from Blackfriars Bridge, in London. According to Hencher, most of his premonitions were preceded by painful headaches – 'a band of steel around my head' – that worsened over the course of days. But his feelings before Aberfan were instant:

It just HIT me without warning. I began to tremble all over the body, felt lethargic and found it very difficult to concentrate on my work. I did mention to a lady sitting next to me, at her enquiry as to whether I was feeling ill, that a big disaster was taking place in this country which could cost many lives. I couldn't and never have been able to pin-point the place of occurrence. (The lady has looked at me in a very worried fashion since, and it seems has avoided being near me).

45

As if he was proposing another chapter for Enoch's volume of rare conditions, Barker posited the existence of what he called a 'pre-disaster syndrome' that might be experienced by a small subset of the population. Barker theorised that some people might have bodily sensations ahead of important or emotional events, not unlike twins who say that they feel each other's pain even when they are hundreds of miles apart. 'Is this perhaps a hitherto unrecognised medical or psychosomatic syndrome akin perhaps to the phenomenon known as the "sympathetic projection of pain"?' Barker asked, in a paper for the journal of the Society for Psychical Research, which was published in 1967. 'Do their symptoms depend upon some sort of telepathic "shock wave" induced by a disaster? But if so, why do they appear to be experienced in advance and not at the time . . .? Clearly these "human

disaster reactors" would appear to require much further study, including perhaps full psychiatric investigation.'

Barker saw great promise in the Aberfan premonitions, but he was also aware of the difficulties that stymied research of this kind. Like most other extraordinary intuitions, these had all been collected after the fact. There was also the metaphysical question of what people might be sensing: was it the disaster itself, or the emotional shock and the grief that it caused? 'The Aberfan disaster cannot be separated from its news,' Barker wrote. And what was the use of this information? Barker acknowledged that even if these dreams and warnings had been publicly recorded at the time, there was no reason why they would have been believed or acted upon: 'Firstly because their premonitions would probably have been insufficiently

clear,' he wrote, 'and secondly because no means existed for them to communicate them to the proper authorities.' Many of the percipients who responded to Fairley's article were simply grateful for being heard. 'These are my facts!' the bank researcher wrote. 'I welcome the opportunity to inform you of these things as most people are inclined to think one mad if one discusses them.'

*

On 29 November, just over a month after the tip slide, the public inquiry into the Aberfan disaster began hearing witnesses at the College of Further Education in Merthyr Tydfil. The college was a low, modernist building, completed in 1952 after slum clearances in the town. The hearings opened on a Tuesday. Across the country, there was heavy fog and snow and ice on the ground. Four surviving teachers from Pantglas Junior School attended, looking young in their suits.

One of the questions posed by the Aberfan tribunal was whether the disaster should have been foreseen. In December 1939, some twenty-five years earlier, almost two hundred thousand tons of coal waste toppled down a hillside five miles south of Aberfan, in the same valley. That tip fell at 1.40 p.m. and buried a five-hundred-foot stretch of the main road to Cardiff. The river Taff was filled to a depth of fifteen feet. By extraordinary good fortune, no one was killed that afternoon but the accident cost £10,000 to repair and prompted the preparation of a report, 'The Sliding of Colliery Rubbish Tips', which was circulated among mining engineers in south Wales and filed away, mostly unread.

Tip number four, above Aberfan, slid for 1800 feet down Merthyr Mountain in 1944. Tip number seven had had its own partial collapse in 1963. The tribunal concluded that three worrying coal slips in the Merthyr Vale had given the colliery ample notice of a possible tragedy, but they also occurred infrequently enough 'to pass unheeded into the limbo of forgotten things'. Until 144 people died in Aberfan, the National Coal Board did not have a procedure for choosing where to put its millions of tons of waste or how to monitor them. During the tribunal, Joseph Baker, the mechanical engineer who picked the site of tip number seven, was called as a witness. Baker was sixty-three and recently retired. He had worked for the mines since he was fourteen years old. He told the tribunal that unless a waste heap started to move, it was considered safe. That was the way it had always been. 'Is it not a very wrong way?' a lawyer asked him. 'Probably, sir,' Baker replied. 'We did not see it.'

After people complained about tip number seven in 1960, Baker walked up the hill and drove some pegs into the ground in front of it, which he visited from time to time. These were gradually covered by slurry, but he did nothing about it. In 1963 and 1964, a local government engineer and a councillor wrote to the National Coal Board, worried about 'The Danger from Coal Slurry being tipped at the rear of the Pantglas Schools'. Aberfan's MP was concerned about the tips but did not want to say anything that might threaten the future of the mine. Nobody thought that anyone would die. In the village, everyone knew that the tip sat on top of a spring because the water course was marked on the local Ordnance Survey map. Children used to play in a pond that it fed. From 1949, people in the village, including the headmistress of Pantglas School and the residents of Moy Road, complained to the mine about how the tips polluted the water coming off the hills and caused floods that were greasy and black with coal waste. The slip in 1963 covered the pond entirely and for the next three years, the toe of the tip crept forward into the valley, 'as though it were being overturned by pressure from behind,' according to one witness. Children and sheep got stuck in the slurry and had to be pulled out.

In the months before the disaster, the slingers working the tip calculated that it moved twenty or thirty feet in relation to a dead tree that they called Hangman's Tree. 'There is a river underneath, that is why it must be sinking,' Leslie Davies, who was in charge of the gang, told his team. At the tribunal, Davies said that it wasn't his job to interpret what he saw on the hill above the village. 'All I was paid for was tipping muck and getting rid of it,' he said. 'I wasn't paid for anything else.'

Signals of the Aberfan disaster were everywhere before it happened, but nobody thought enough, or feared enough, to connect them and to see it coming. Among the tribunal's conclusions:

17. We found that many witnesses, not excluding those who were intelligent and anxious to assist us, had been oblivious to what lay before their eyes. It did not enter their consciousness. They were like moles being asked about the habits of birds.

While the hearings were going on, an American photographer for *Life* magazine, Chuck Rapoport, arrived in Aberfan to record its grief. Rapoport was twenty-nine and had a son, who was six months old. He took a room at the top of the Mackintosh Hotel on Moy Road, where the first emergency call had gone out. From his window, he looked out on the devastation. The village had been overrun by reporters and enquiring visitors of all kinds. 'Like wild animals pillaging our souls,' an old man told Rapoport, on the day that he arrived. The Queen had just been through.

Rapoport had never been to Wales. His last essay for *Life* had been with detectives in central Manhattan, shooting hookers and other marginal characters in Times Square. 'My idea of Wales was of a dark, grim, rainy, depressing place and that's exactly what I found when I got there,' he said. 'It was the beginning of winter. It snowed while I was there. It rained a lot while I was there. The slurry from the mountain that came down was still apparent on the streets, piles of it.' Rapoport bought a pair of rubber boots, which he rarely took off. His room was terribly cold; he acquired extra heaters, which fused the electricity in the rest of the building. He spent a lot of time in the pub downstairs, where the

men of Aberfan came to talk and joke of other things. When Rapoport encountered men on their own, they would often weep. 'They knew I was a stranger and I would go away,' he said.

The photographer noticed that the people in the village fell into two groups, those who said the disaster was an unimaginable event, and those who claimed that it had been entirely predictable and that they would never forgive themselves for not stopping it. 'There were guys in the club . . . They would say, "Chuck, could you ever think that something like this was going to happen and all our kiddies would die?" They would go on like this, like they were so astounded,' Rapoport recalled. 'That was one group. And then the other group would say, "This was bound to happen. You should come here in the spring when the water is coming out from under the tip; you could just see the water coming out. What's holding this tip?" They knew it.'

While he was in Aberfan, Rapoport also heard the story of Eryl Mai Jones, the girl who dreamed that something black had come down over the school before dying in it herself. He spent time with Eryl's mother, Megan. The family had an ironmonger's shop. People in the village interpreted the child's vision according to their own conception of the disaster. Rapoport didn't know what to think. It could have been a prophetic dream; it could have been the kind of nonsense that comes out of a child's mouth on any given morning. 'I can understand not believing that kind of stuff,' Rapoport said. 'You don't know whether that kid actually had a premonition that something bad was going to happen, or whether she was just trying to get out of school.'

Rapoport's photographs from Aberfan are searing. Grief errs on madness. While he was there, he heard about a man who had lost everything. John Collins was an engineering inspector, who was at work in Cardiff when the slide destroyed his house on Moy Road, killing his wife and two sons: Peter, who was at home, and Raymond, who was walking to the senior school. All that he had was gone. About a month afterwards, Rapoport and John Hicks, a *Life* reporter, met Collins in the front room of his sister-in-law's house. Collins was wearing a borrowed suit – he had no clothes – and smoking a cigarette.

Hicks interviewed the stricken man. But when the moment came for Rapoport to take his portrait, the photographer froze. 'For the only time I was there, I was immobilised by his grief,' he said. 'I just felt it was so distasteful to raise my camera.' But Collins encouraged him. 'Go on, man,' he said. 'It's your job.' Rapoport shot a roll of film. The pictures are harrowing, off-kilter. Leaves spiral up the wallpaper. Someone else's wedding picture is on the bureau. Collins covers his face. Rapoport's photo essay was published in *Life* in February 1967. An American linguist living in Brussels bought the magazine. She was separating from her husband, a Czech athlete who had defected to the West. The woman was Catholic and her mother came from Newport, in Wales. The linguist saw Collins' portrait in *Life* and wrote to him. They were married for twenty-two years and had a daughter, whom they called Bernice. Rapoport's photograph made a family.

*

How do you account for the role of chance in your own life? When we were getting married, my wife and I chose the symbol of two magpies, for joy, to put on our invitations. We became quite focused on magpies and reported our sightings to one another. My wife was also pregnant. Early one morning, a few weeks before our wedding, heady with chores and expectations for the life that we were making, we looked out of our bedroom window and there were three magpies, for a girl, quietly hopping about in the garden. We never asked for a test to confirm the sex of our daughter because we felt we had already been informed. It can be very difficult, even in the moment that something noticeable is happening, to separate an event from the meaning that we choose to give it. With time, once an unlikely occurrence is incorporated into the story of a life, or a death, it becomes almost impossible to see the alternative possibilities that once existed. Storytelling itself is also an act of narrowing, of eliminating branching futures, until there is only one way that things were ever going to go. We confer meaning to control our existence. It makes life liveable. The alternative is frightening. Randomness is banal. It diminishes us. But the truth is that we resist meaning almost as often as we insist upon it. We refuse its presence to make life simpler and to spare

ourselves. There was no way we could have seen that coming. We didn't stand a chance. It is easier to be a mole who knows nothing about the habits of birds. Letting things go, surrendering to chance, is its own narrative act but we talk about it much less. The magpies that alight in the garden and go uncounted; the visions that we wave away; the associations that we refuse to acknowledge; the tragedies that were unstoppable. How we distinguish the chances that signify and the ones that do not, and the decisions that we make in our lives as a result, is the kind of question that turns in on itself and might be impossible for us, as individuals, to answer. We cannot stand outside our own lives. We would not want to.

Most of us make some kind of untidy peace with these questions, if we think about them at all. But a few people find it very difficult to leave chance alone. In December 1931, Arthur Koestler, a young science journalist in Berlin, decided to change his life after a shattering night during which he lost a fortune at cards, slept with someone he didn't like and wrecked his car. He joined the Communist Party a few days later and wrote *Darkness at Noon*, a classic novel about totalitarian power. After the war, Koestler settled in Britain. During the sixties, when Barker became occupied by similar questions, Koestler remained haunted by the meaning of coincidence, particularly by odd events that seemed to cluster together. 'When major and minor calamities crowd together in a short span of time, they seem to express a symbolic warning, as if some mute power were tugging at your sleeve,' he recalled in his autobiography, *Arrow in the Blue*, about that decisive night in Berlin. 'It is then up to you to decipher the meaning of the inchoate message. If you ignore it, nothing at all will probably happen; but you may have missed a chance to remake your life, have passed a potential turning point.'

*

Barker was determined to broaden his Aberfan experiment. Fairley's article had been followed up by other newspapers. The Institute of Psychophysical Research, in Oxford, had put out its own request for premonitions of the disaster and collected two hundred. Barker considered writing a book about the subject. He worked long days at the hospital and then came home and worked in his ground-floor study at Barnfield, the house where he lived with his wife, Jane, and their three children, outside the village of Yockleton, on a quiet road that led to Wales.

Barker had met Jane Homfray at St George's Medical School, in London, in 1946. He was studying to be a doctor and she was training to be a nurse. The Homfrays were from Gloucestershire; the men in Jane's family served in the military or held posts in the colonies. Her father had been a district officer in Nigeria and died in a hospital for tropical diseases when she was seven years old. Jane grew up with her mother and two younger siblings in a cottage not far from Cheltenham. She had brown hair, a wide mouth and proper

pronunciation. Jane's medical training meant that she kept up with Barker's research; she stayed up late talking with his colleagues when they came over. Jane contributed an ink drawing of a stag and a cherry tree to illustrate a story from the adventures of Baron Munchausen in Barker's doctoral thesis about the syndrome. She teased Barker; he made her laugh. In the winter of 1966, Jane was pregnant with their fourth child. She loved their busy, growing household. Barn-field, which they rented, was a large, semi-detached villa with a roomy garden that gave on to a neighbouring farm. The children, Nigel, Josephine and Julian, had space to run. Their neighbour made violins in a workshop on the property.

Barker's sudden departure from Herrison Hospital, three years earlier, had put a strain on the marriage. But his life and career had recovered and stabilised in Shropshire. He found the rural peace helpful for writing. Beyond his premonitions project, his more orthodox medical research was also catching attention. In September, a few weeks before the Aberfan disaster, a BBC television crew had spent three days at Shelton, filming his work with gambling addicts, for the science programme *Tomorrow's World*. In December, another aversion therapy case that Barker and Miller described for *Pulse*, a medical magazine, made headlines around the world. Barker and Miller reported that they had cured a thirty-three-year-old married man, whom they called 'Mr X', of an extramarital affair. The man had fallen in love with his neighbour and his wife had tried to drown herself in the bath. On a darkened ward at Shelton, Barker and Miller showed Mr X alternating pictures of his lover and his wife and treated him with half-hour sessions of electric shocks, measured at seventy volts and administered to his wrist. 'Immediately after the first session, he developed a deep sense of guilt and broke down completely,' they wrote. 'Subsequent sessions appeared less traumatic but nevertheless left a deep impression on him.' After six sessions, Miller and Barker described the man as 'completely indifferent to his former lover'. They identified infidelity as 'a common and fascinating problem' they were eager to explore further.

The story ran in newspapers across America. 'Psychiatrists Pull Switch', reported the *Sacramento Bee*. 'This may be good news for Mr and Mrs X, but the futuristic implication that the commonplace old love triangle can now be eliminated by an electrician is depressing,' wrote Russell Baker, a columnist at the *New York Times*. At a time of great

61

progress and experimentation, Baker imagined humans being blandly reprogrammed by drugs and shocks to want what was good for them: 'It could be used to make adolescents want, once again, to come home on Saturday nights; to make mothers want, once again, to do the dishwashing; to make fathers, once again, give up their TV football and go back to their families.'

When it came to Aberfan, Barker was keen to attract as much attention as possible. As a result of his newspaper scoops, Fairley was a regular science commentator on the BBC and on ITV, Britain's first commercial TV channel. The two worked together to publicise the premonitions. On 2 December, five weeks after the first appeal appeared in the *Evening Standard*, Fairley, Barker and a number of the Aberfan percipients were invited to appear on *The Frost Programme*, a live ITV interview show with David Frost, the twenty-seven-year-old star of late-night television. The show was broadcast three nights a week, after the ten o'clock news. It was Frost's first attempt at serious journalism after making his name as a satirist in the early years of the decade. He and his team worked at Television House in Holborn during the day and often lunched at L'Escargot in Soho before driving out to the Rediffusion TV studios in Wembley Park, in the London suburbs, for the evening. Miss Middleton and Grace Richardson, the film technician who smelled death, were among around a dozen of Barker's seers who were invited to the broadcast. Some travelled hundreds of miles to take part.

The night of the broadcast was the first time Fairley and Barker had encountered most of the Aberfan percipients in person. When they gathered in the green room, Fairley was taken aback. '"Weirdos" would be too strong a description, but they were certainly "different",' the journalist later wrote. About twenty minutes before the programme was due to start, Frost came into the room, made some light conversation and then disappeared. During the first half of the show, Frost interviewed John Betjeman, the poet laureate. Barker and his seers were supposed to appear after the commercial break. On a monitor, the group watched Frost in conversation with the production team. After the break, he continued talking to Betjeman. The call never came. In the end, the entire programme became an impromptu, extended audience with Betjeman, who led the studio audience in reciting their favourite poems. 'The forty minutes raced by in what seemed like no time at all,' Frost remembered later.

After the broadcast, Frost came backstage to apologise. Barker was still furious when he got home to Shrewsbury. He had told Enoch that he was travelling to London to appear on the programme, but not why. 'He was very, very, very cross,' Enoch recalled. But after the encounter in the green room, Fairley understood Frost's reluctance to allow the group on national television. Aberfan was still raw in people's minds. The seers' visions were fragmentary and easy to pick apart. Fairley suggested to Barker that they take a more ambitious and open-ended approach: to log premonitions as they occurred and see how many were borne

63

out in reality. 'The world is full of people who claim to have seen something coming but they always speak out after the event,' Fairley wrote.

In the weeks before Christmas, Fairley and Barker approached Charles Wintour, the editor of the *Evening Standard*, to open what they called a Premonitions Bureau. For a year, readers of the newspaper would be invited to send in their dreams and forebodings, which would be collated and then compared with actual happenings around the world. Wintour was a stylish, sophisticated editor. He had joined the *Standard* as an opinion writer in 1946. Old-timers said that he changed abruptly after his eldest son, Gerald, was killed in a traffic accident in the late fifties. He gained the nickname 'Chilly Charlie' and since 1959 he had transformed the *Standard* into London's upmarket evening paper. Wintour hired young writers and editors and oversaw them exactingly, sending back copy and short, acerbic notes of congratulation. (Brave souls occasionally mimicked these and pinned them up around the newsroom.) When Wintour was really pleased, he was known to tap the desk with his middle finger.

The heart of Wintour's paper was the diary desk and the features section, which sat outside his office. If he were at a loose end, he might take a promising writer to the Savoy for lunch. Most specialist reporters, like Fairley, were part of the news operation and less interesting to him for that reason. But Fairley was an unusual case. He covered space. He had sharp instincts, which had proved, on occasion, to be preternatural. His stories were popular with younger readers, and he had a talent for showmanship. Like other great newspaper editors, Wintour was game for unusual ideas. He agreed to the experiment. Fairley had a date stamp made for the Premonitions Bureau. He devised an eleven-point scoring system for the predictions: five points for unusualness, five points for accuracy, and one point for timing.

*

Barker and Fairley prepared to start logging premonitions in the first week of 1967. As Christmas approached, the *Evening Standard*, like most of the country's major newspapers, had a reporter standing by to cover Donald Campbell's attempt to break the water speed record on Lake Coniston, in the Lake District. Campbell was an idol of Britain's post-war jet age. He chased speed records on land and on water in a series of vehicles all named *Bluebird*, after the Maeterlinck play. He compared the urge to achieve greater velocities to exploring. 'The faster man travels the more difficulties he encounters, the more he is determined to overcome and understand them; and as he proceeds, stage by stage, he penetrates farther into the unknown,' Campbell wrote in 1955. 'It becomes something of a disease in the blood, which feeds on inclination and atmosphere.'

By the late sixties, Campbell was an antique sort of hero. He used powerful, experimental technology; he was also strongly superstitious. An enamel medallion of St Christopher, the protector of travellers, was screwed into his instrument panel. He carried Mr Whoppit, a lucky teddy bear, every time he climbed into a cockpit. He loathed the colour green. On Lake Coniston in the winter of 1966, Campbell named his fears aloud and confronted them anyway. On 13 December, a bright, frosty day when no one expected him to take the boat out, Campbell piloted *Bluebird K7*, his jet-engined hydroplane, up to 267 mph and hit a seagull, which he considered a bad omen. The collision made a dent on the boat, which he refused to fix. He told a television crew about the time when he had driven his gas turbine-powered car over 400 mph on damp, treacherous sand on Lake Eyre in Australia in 1964. Campbell had been afraid to turn the car round and complete his record attempt. While he sat still in the desert, an image of his father, who had also been a speed record breaker and who had died in 1946, appeared as a reflection in his windscreen. 'Don't worry. It'll be all right, boy,' his father said, and Campbell drove back even faster than before. 'Explain it as you will – I cannot. But it happened,' he told rapt reporters on the lake shore.

On Christmas Day, with no engineers or safety team, Campbell persuaded a friend in the village to help him take *Bluebird* out on the water and he roared up and down alone. At a New Year's Eve party at the Sun pub, he toasted the press at midnight. 'I know that you are all waiting for me to break my neck,' he said. Campbell played cards to pass the time, waiting for the lake to still. A few evenings later, after a day of sleet and frost, Campbell was playing Russian patience while he waited for a card game to assemble at his bungalow. He dealt himself the ace of spades, followed by the queen. He told David

Benson, a friend who wrote for the *Daily Express*, that Mary Queen of Scots had drawn the same cards before her beheading in 1587. He stayed up late. 'I have the most awful premonition I'm going to get the chop this time,' Benson remembered Campbell saying. 'I've had the feeling for days.'

The next morning was 4 January, a Wednesday. Campbell had a breakfast of cornflakes and a coffee, laced with brandy. There was a slight swell on Lake Coniston but it was calm enough to launch *Bluebird* at 8.40 a.m. In order to break his own water speed record, Campbell had to complete two one-kilometre runs – up and down the lake – at an average speed of more than 276.33 mph.

At 8.50 a.m., the first edition of the *Evening Standard* went to the presses, announcing the launch of the Premonitions Bureau. 'If you dream of disaster . . .' ran the article's headline. At the same minute, Campbell entered the second one-kilometre run of his water speed attempt on Lake Coniston at 328 mph. He was beyond the world record, well into the unknown. He had not left enough time for the wake of the hydroplane to settle on the lake and as Campbell sped back, *Bluebird* began to bounce hard on the water. She rose high into the air, somersaulted and killed him. Photographs of the flying boat and the story of Campbell's ominous cards filled the front page of the newspaper by the late afternoon. A radio recording preserves Campbell's last words as he streaked along. 'Hello, the bow's up . . . I'm going,' he says. And then there is the sound of a small sigh.

II

Premonitions are impossible, and they come true all the time. The second law of thermodynamics says it can't happen, but you think of your mother a second before she calls. There is no way for us to see, or feel, things before they occur but they often seem to hang around regardless. Chance meetings, friends, lovers and deaths prefigure in our minds. In John Berger's novel *G.*, the narrator, while shaving, thinks of a friend who lives in Madrid and wonders if he would still recognise him in the street, fifteen years after their last meeting. Then he goes downstairs and finds a long letter from his friend in his mailbox:

> Such 'coincidences' are not uncommon and everyone is now more or less familiar with them. They offer us an insight into how approximate and arbitrary is our normal reading of time. Calendars and clocks are our inadequate inventions. The structure of our minds is such that the true nature of time usually escapes us. Yet we know there is a mystery. Like a never-seen object in the dark, we can feel our way over some of its surfaces. But we have not identified it.

Witnessing the future was more common in the past. The Bible hums with foretelling. In the Book of Samuel, we learn that before prophets were prophets they were known as *ro'eh* or *ho'zeh* – from the ordinary Hebrew verb to see. 'And your sons and your daughters shall prophesy, and your young men shall see visions, and your old men shall dream dreams,' the Lord says, in Joel. I had never come across that quote, or heard of the Book of Joel, until I read it in the dedication to Miss Middleton's memoir in November 2019. The following morning, I walked into my bedroom and heard it on the radio.

The rational explanation for premonitions is that they are coincidences. But it is not easy for us to accept this. Our brains argue against it. As creatures, we prefer patterns to none. In the late eighteenth century, Immanuel Kant proposed that instead of passively absorbing reality and taking things as they come, our minds are much more active and constructive – we infer and imagine, shape and constrain our perceptions even as we have them. 'Objects', he wrote in 1787, 'must conform to our cognition.'

Kant was sure that he was right: that it is our minds that make the world, rather than the other way around. He compared himself to Copernicus, who proved that the Earth was moving, rather than the sun. But Kant's psychology of perception was ornate and hard to prove. Many philosophers found it slightly embarrassing. In the mid-nineteenth century, however, the German polymath Hermann von Helmholtz argued that Kant was correct when it came to the eye. In his *Handbook of Physiological Optics*, Helmholtz suggested that much of our vision is made up of 'unconscious inferences' about what we expect to see, rather than a straightforward processing of light and shapes.

Concepts such as space and time, which exist inside our heads, help to organise an otherwise indecipherable array of partial, flipped images that flash across our retinas. In a lecture that he gave in 1855, Helmholtz recalled the moment that he became aware of perspective, as a small boy in Potsdam. 'I was taken past a high tower, on the topmost gallery of which people were standing, and begged my mother to lift down the little puppets,' he said. It had not dawned on him until then that things which are further away are smaller. Helmholtz noted that these inferences, once acquired, reside solely in our brains but have the power to shape reality itself. Once seen, they cannot be unseen. He never saw the puppets at the top of the tower again. 'These qualities of sensations belong only to our nervous system and do not extend at all into the space around us,' Helmholtz wrote in 1878. 'Even when we know this, however, the illusion does not cease, for it is the primary and fundamental truth.'

Relying on these illusions may be the only way that we can cope with the storm of information that reaches us. It has obvious evolutionary advantages. Extrapolating from fragments of information and relying on memories, allows us to move faster through the world and avoid mishaps along the way. It is safer to predict what we are seeing (is that a tiger in the shadows?) than to wait and come to some conclusions later. We internalise concepts – chair, dog, bird –

from the slightest hints. An infant does not need to memorise 350 possible breeds of dog in order to tell apart a dachshund from a squirrel in the park.

Since the 1990s, Helmholtz has been cited by neuro-scientists as the forefather of 'predictive processing', the idea that our entire brains work this way. In the manner of Kant, this theory of perception overturns the classical model of experience. Rather than sensing the world in a 'bottom-up' fashion, through our eyes and ears, a ripple on your skin, it proposes that our brains work from the top down, a waterfall of internal theories and beliefs, memories and expectations, which guide our perceptions and are then corrected by feed-back from the outside world. When we walk into our kitchen, we only see what our brains have not already placed there. Is that a fox in the sink? These surprises are called 'predic-tion errors' and our brains work hard to eliminate them – the glitches between our illusions – and to generate new explana-tions of the world. 'It's utterly seamless. I think it's the thing that underwrites our conscious experience,' Phil Corlett, a professor of psychiatry at Yale, explained.

Some of the best evidence for the idea that we are always predicting, rather than merely perceiving, comes from the brain's mistakes, when we come up with faulty accounts of reality. In one well-known experiment, researchers put a different image in front of each eye at the same time, like a face and a house, and study how the brain resolves the situ-ation. One of the inferences (neuroscientists now call these 'priors') that we hold dear is that only one object can be in the same place at the same time. So instead of melding the objects under each eye into something novel and strange, our minds tend to see one after the other. Face. House. Face. House. The images hover in and out of focus, in an unstable fashion, until a more comprehensible world resumes.

More serious delusions, such as the hallucinations and paranoia which are symptoms of schizophrenia, can also be explained by the model of the predictive brain. The single most influential proponent of the theory, a neuro-imaging expert named Karl Friston, worked as a psychiatrist in a mental hospital outside Oxford in the 1980s. One of his patients was obsessed by angel shit. Friston was impressed that a mind could become occupied by such a question. According to Friston, such delusions develop when, for one reason or another, the relationship between our expectations and the feedback from the world goes awry. We fail to correct our hypothesis, even though it is wrong. She loves me. Or we respond too strongly to stimuli, seeing meaning where there is none. The shadow that looks like a tiger becomes a tiger. Why am I covered in angel shit again?

The cascade of beliefs, memories and theories that form our predictions takes place in the frontal lobes of the brain, which are much larger in humans than in other animals. Diseases or injuries which damage that part of the brain can leave people unable to sense the future, or to consider the possible consequences of their actions. The first sign that someone is suffering from the behavioural variant of fronto-temporal dementia, also known as Pick's disease, is often that they have gone slightly mad. They ask intimate questions of strangers. They strip off all their clothes because they are hot. They insult their bosses in meetings. They spend money like there is no tomorrow because they have become unshored from tomorrow. In 2015, the *New England Journal of Medicine* reported on a patient known as 'Case 9', with frontotemporal dementia, whose first symptom was that he refused to let his wife listen to his iPod. A previously garrulous and sociable man, he stopped wanting to talk to people. He ate and drank until he was sick. Things stopped following, one from another. He played the same audiobooks on repeat. When his wife went into labour, preparing to give birth to their first child, she had to ask him to stop listening to *Harry Potter* on his headphones. When we stop seeing where things are going, we cease to be ourselves. It is human to think ahead. Premonitions are tantalising because they are simulacra of this essential mode of thinking. What is the difference between an impossible hypothesis of the world and an insight that no one else has managed to see yet? 'We get excited by things that appear to give us a leg up in the predictive world because that's really, really advantageous,' Corlett said. 'It's sort of what we were put on the earth to do.' A more predictable existence is, in theory anyway, a less frightening one. Societies have always craved prophets, who claim to see round the next corner.

*

The BBC Home Service broadcast an item about the Premonitions Bureau the morning after Campbell's death on Lake Coniston. The *Today* programme reported that 'this sort of foresight is now being taken a good deal more seriously'. During an interview with Fairley, which went out just before 7.20 a.m., millions of people in Britain heard how the *Evening Standard* had collected seventy apparent premonitions of the Aberfan disaster and was now embarking on a year-long experiment to investigate the phenomenon more widely.

'We're asking anyone who has a dream or a vision or an intensely strong feeling of discomfort which seems to involve somebody else, danger to somebody else, or to themselves, to ring us,' Fairley said. He gave out the newspaper's switchboard number: Fleet Street 3000. 'We will log this information and look at it very carefully and we have a small team of investigators who will attempt to corroborate whether the premonition comes true,' Fairley went on. He was asked whether he was expecting lots of visions of certain events. 'Well, I have a horrible feeling that we should get a lot of letters and phone calls from cranks and highly imaginative people,' he replied. 'But I don't want to put these people off necessarily because a lot of highly imaginative people are sometimes thought to be the people who genuinely have a gift of premonition.'

The Premonitions Bureau became a department within Fairley's small, cluttered domain on the long row of desks for specialist reporters in the centre of the *Evening Standard* newsroom. After twelve years at the newspaper, and with the title of Science Correspondent, which he had given himself, Fairley had his own desk, book cupboard, a few filing cabinets and a microfiche viewer. He also had an assistant, Jennifer Preston, who had joined the newspaper a few months earlier from the *Evening News*, the *Standard*'s rival. The *Evening News* sold more copies than the *Standard*, but it was more downmarket – all murders and racing results. Preston, who was thirty, had grown up in Elmers End, a quiet south London suburb, on the border with Kent. She had two young sons and lived with her husband Michael, who sold television rentals and later drove a minicab, in a one-bedroom flat in West Grove, a dilapidated manor house overlooking Blackheath that was sometimes used as a set in horror films.

Preston was a striking figure on the newsroom floor. She had black hair, high cheekbones and a roman nose. At the *Evening News*, she had been the assistant to the features editor. She helped reporters with their research, carried out interviews and kept on top of errands and loose ends. She had a way of being involved in a dozen things at once but was personally contained. 'She would have been a very good captain in the army,' her colleague Bob Trevor recalled. Preston was an autodidact, with a roving curiosity. She read Latin easily, was fascinated by plants and the ancient world, and was a keen follower of county cricket. Later in her life, when the brickwork of the family house needed mending, Preston hired some scaffolding and did it herself. She exchanged letters with President François Mitterrand of France as if that were a normal thing to do.

'She had that "Just get up and do it" mentality. There wasn't any dilly-dallying,' her daughter, Arabella, said. 'She didn't do self pity at all.' Preston was a natural fit for the Premonitions Bureau. She had an abiding interest in the occult. She couldn't pass a gypsy selling lavender without falling into conversation.

Barker corresponded with the percipients from his Aberfan study but members of the public who got in touch with the Premonitions Bureau reached Fairley's desk at the *Standard*. Most of the time he wasn't there, so it was Preston who logged the telephone calls, filed the letters and cross-referenced the visions that arrived against a dozen daily newspapers, checking for possible matches. She sorted the bureau's warnings into fourteen categories, which included 'Royalty', 'Personalities', 'Racing', 'Fire' and

'Non-specified disasters'. Each message sent to the bureau received a pro forma reply:

> Mr Peter Fairley, Science Correspondent of the *Evening Standard* thanks you for your courtesy in writing to him about premonitions. The material you provided has been entered in his records. He hopes that if you have further premonitions during 1967 you will let him know.

Fairley liked to gamble. He went on betting sprees on the horses, fired by odd names and numbers that lodged in his mind – a mental process which he believed to be a form of premonition. On quiet mornings, he enjoyed going through Preston's racing category for possible tips.

In print, Fairley professed a careful neutrality about whether it was possible to see the future. 'I make only two promises,' he wrote in the *Evening Standard* during the bureau's first week in operation. 'Nobody will be scoffed at. And their premonitions will be treated as confidential until the inquiry is complete. Let us simply get at the truth.' Privately, Fairley had his own theories about how precognition might work. He wondered if people glimpsed the future through a form of telepathy. He compared our thoughts to radio waves, which other minds were able, intermittently, to tune into. He thought that if the phenomenon was real, it was likely to be almost entirely subconscious or out of the control of the percipient.

But Fairley's profession, and his gift, was as a populariser. 'Man has been into space and come back alive,'

Fairley wrote on the morning of Gagarin's flight in 1961. 'He zoomed gloriously into orbit atop a mighty Russian rocket. Buffeted. Deafened. Contorted – but safe.' Fairley narrated the scientific progress of the decade in a tone that was heroic, breathless, tinged with risk. He was fond of explanations that involved everyday objects; Gagarin's rocket weighed the same as sixty red London buses. He coined the phrase 'brain drain' to describe the loss of British scientific talent from the country's top universities and companies. In 1962, Fairley was tipped off ahead of a landmark paper in *Nature*, based on measurements from radio telescopes in Cambridge, that upheld the Big Bang theory of the formation of the universe. The *Evening Standard* sent a newspaper seller, dressed in a white coat, to hawk copies of Fairley's front-page story to his rivals as they trooped into a press conference to hear the findings. From week to week, Fairley used his 'World of Science' column to declare the arrival of various ages: the computer age, the electronic age, the space age. He hailed lasers, atomic rockets, superconductivity and high-pressure physics. He wrote a book about pain. 'Your reputation is expanding as fast as the Universe,' Wintour wrote to him after his Big Bang scoop. Fairley's television appearances, on the BBC and ITN, became more frequent as the space race intensified. He began to be recognised in the street. In his own mind, he addressed all his stories to a single imaginary reader: a working-class woman in a brown dress, who was married to a lorry driver and lived in Wapping, a run-down neighbourhood on the banks of the Thames in east

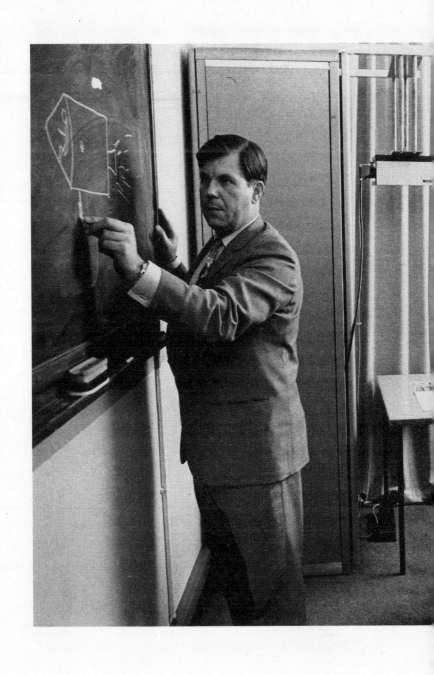

London, a figure that Fairley realised later in his life was based on his grandmother.

In the form of telephone messages, careful letters and uncertain, scribbled notes, the warnings of the Premonitions Bureau joined the cacophonous, paper-strewn world of the *Evening Standard* newsroom. The broad second-floor office thrummed with a hundred typewriters, telephones ringing without end, chairs scraping on linoleum, the whistle of an overhead pulley system yanking photographs from one desk to the next, the floor shaking nine times a day as the printing presses heaved in the basement, shouts of 'Boy!' as messengers, who waited against the wall, were sent on errands throughout the building and beyond. Regardless of the season, the air in the office was lit from morning until night by fluorescent tubes and heavy with cigarette smoke, bad breath and casual sexism. People worked with a feverish, intermittent focus that came and went as the editions closed throughout the day. They laughed and larked about one minute and typed irritably the next. The laconic, perceptive figure of Wintour slipped between the desks. When people started working on the second floor at Shoe Lane, they tended to get terrible headaches for the first six weeks. Then they found it difficult to work anywhere else.

The Premonitions Bureau received twenty warnings in its first forty-eight hours. One spoke of a train crash. Two predicted passenger planes coming down in the Atlantic. Another said a ceiling would fall down at a department store on Kensington High Street called John Barker & Company.

(A tug on the sleeve, given that Barker's name had not appeared in the newspaper.) 'We shall see . . .' Fairley wrote in the *Standard* on 6 January. If the first year of the experiment showed potential, Fairley intended to present the results to Parliament and to the Medical Research Council, to see if they justified an official national early warning system of some kind. On the BBC, Fairley was asked what he would do if the bureau recorded, say, fifteen similar premonitions of a looming disaster. 'Clearly if one had remarkable similarities and a large number of premonitions affecting a specific event,' he said, 'I couldn't possibly stand by.'

*

Sigmund Freud kept a copy of Helmholtz's *Handbook of Physiological Optics* on a shelf above his couch. Like Kant, he also proposed a model of perception that seems to foreshadow the concept of the predictive brain. Freud described an interplay between the primary, untethered instincts and desires of the mind, which he named the id, and the ego, which negotiated the encounter of those desires with the real world. Synthesising the two – our inborn expectations and the lives we actually lead; the pleasure principle and the reality principle – was the fundamental task of the healthy mind. To do so with as little fuss as possible (minimising our prediction errors, as neuroscientists would say) was something that Freud sometimes called the Nirvana state.

He could never quite make up his mind about the occult. In April 1909, Freud had an argument about precognition with Carl Jung, who was much more sympathetic to the concept. The analysts were in Freud's apartment in Vienna. As the two men disagreed, Jung experienced an odd hot sensation in his chest – 'as if my diaphragm were made of iron and were becoming red-hot – a glowing vault.' Then a crashing sound came from a bookcase. Both men stood up in surprise. Jung attributed the disturbance to extra-sensory perception, or ESP. 'Oh, come,' Freud replied. 'That is sheer bosh.' But he had a good look afterwards.

Outwardly, Freud worried that psychoanalysis, with its focus on the mysteries of the unconscious, explored similar territory to the supernatural and could end up being terribly harmed if he appeared too sympathetic to doubtful science. He dreaded what would happen if a single occult phenomenon was found to be real. 'There may follow a fearful collapse of critical thought, of determinist standards and of mechanistic science,' he wrote in 1924. Nonetheless, Freud spent a lot of time thinking about chance. Psychoanalysts are not keen on randomness. They like meaning to come in layers. But a world without accidents is also a world of hidden strings and predestined lives. To make room for chaos, and to disavow the supernatural, Freud proposed a difference between what he called the *Unfall*, the interpretable accident, and the *Zufall*, pure, inexplicable happenings.

But he could never quite get past telepathy, which appeared to constitute communication between unconscious minds, a phenomenon that Freud believed he had experienced with his patients. As it did for many thinkers who lived through the technological advances of the nineteenth century, during which the telegram and the telephone brought voices and messages, like apparitions, across previously uncrossable distances, for Freud telepathy existed merely as a problem to be solved. Like Barker, he was a member of the British Society for Psychical Research. He joined the group in 1911. 'I do not belong with those who reject in advance the study of so-called occult phenomena as being unscientific, or unworthy, or harmful,' he wrote to an English psychical researcher, Hereward Carrington, ten years later. 'If I were at the beginning of my scientific career, instead of at the end of it, as I am now, I might perhaps choose no other field of study – in spite of all its difficulties.' Freud's moments of weakness towards the occult unnerved many of his closest disciples, who tried to explain them away or begged him to remember himself. Ernest Jones, Freud's long-time friend and biographer, was very upset by the whole business. In 1926, Freud tried (and failed) to reassure Jones in a letter. 'Just answer calmly that my acceptance of telepathy is my own affair,' he suggested, 'like my Jewishness and my passion for smoking.'

*

John Barker grew up in a society in which one set of certainties had yet to be eclipsed by another. His father, Charlie, was an accountant, whom Barker described as 'a precise, matter-of-fact man'. Charlie was educated at St Lawrence College, a boarding school in Ramsgate, on the south coast of England. When the First World War broke out, he was twenty-three. Charlie volunteered for the Auxiliary Service Corps, the transport wing of the British army. He arrived in France in January 1915 and spent the next three years serving in mechanised units, shepherding lorries, ammunition and food through bombardments and mud up to soldiers in the front lines.

Charlie enlisted as a private and left the army as a captain. He was mentioned twice in despatches, for bravery. Like many soldiers on the Western Front, he had supernatural experiences during the war. The modernity and mass death of the fighting in northern France made it a place of visions and strange happenings. Soldiers on all sides saw crosses in the sky and heard voices that saved their lives. Sensing your death, or that of a comrade in the line, was one of the most common phenomena. British soldiers spoke of it as 'the Call'. Each army had its own omens. Among French troops, it was bad luck to dream of a bus. Often these premonitions didn't materialise. 'I had known quite a few who at various times had such presentiments and I also knew that they were as often wrong as right,' a Canadian infantryman named Charles Savage noted. 'But statistics are poor consolation when you feel that way.' In 1917, the official French *Bulletin des armées de la République* asked soldiers to send in their premonitions so they could be studied.

As a boy, Barker listened to Charlie's occult war stories, but he did not record what they were. The best known apparition among British soldiers on the battlefield was the 'Angel of Mons', the visitation of a line of shining, spectral figures, said to resemble bowmen, which materialised during the retreat at the Battle of Mons on 23 August 1914. The archers were first described in a piece of fiction published in the *Evening News* a month after the battle. The story was subsequently reprinted in religious pamphlets around the country and taken to be real. By the spring of 1915, there were soldiers in Belgium who swore they had seen the angels too.

Back in Britain, spiritualists and mediums offered families the chance to speak to their sons who had been lost in the trenches. Prominent thinkers such as Arthur Conan Doyle and Oliver Lodge, a physicist and pioneer of radio, published heart-rending accounts of conversations with their dead boys. Rudyard Kipling eschewed it all. 'I have seen too much evil and sorrow and wreck of good minds,' he wrote of the lure of the psychical realm. A few years after the war, Kipling dreamed that he was standing in his best clothes in a great hall, on a floor of stone slabs with rough joints. He had the sense of a crowd. He was at some kind of ceremony, but he couldn't see what was happening because of the fat stomach of the man on his left. On 19 October 1926, Kipling was in Westminster Abbey, at the unveiling of a plaque to the million war dead, for which he had written the inscription. It was a solemn, formal occasion. The abbey was full of dark-suited men. But Kipling's view was obstructed by the large frame of the man next to him. Kipling looked around and then down and, in silence, recognised the stone floor from his dream. 'But here is where I have been,' he wrote later. 'How, and why, had I been shown an unreleased roll of my life-film?'

Barker was an especially only child. Charlie was one of four siblings but the only one to ever marry, or leave home. Throughout Barker's life, his uncles, Theodore and Arthur, and his aunt Adelina lived together in a large house on Shooter's Hill, in Blackheath, which had belonged to their parents. Theodore tinkered with cars; Adelina painted. They were solemn, energetic Christians. When Charlie left the army, he got a job as an accountant and manager at a car dealership in Bromley and married Norah Hyne, the daughter of a clergyman from Bedfordshire. The couple lived in a succession of comfortable suburban villas while their son went to Bickley Hall, a local preparatory school. An early snapshot shows him playing cricket with his parents – a set of figures on a broad, desolate plain.

In May 1938, Barker won a scholarship to Tonbridge, a boys' boarding school in Kent. He was a tall, athletic boy. He played rugby and won two-hundred-yard dashes. Barker was fifteen when the Second World War began. Charlie re-enlisted in the army and was posted to Belfast. He took his family with him. Barker enrolled to study medicine at Queen's University in the city in 1941. A series of photographs, taken that September, show him in a dark jacket and tie. He has a soft mouth that bunches together when he smiles. He excelled in Belfast and transferred to Cambridge University in the spring of 1943, where his tutors found him keen and hard-working. 'Still immature, good judgement, will be first class,' his biochemistry supervisor reported at the end of his second year. Barker was the secretary of the rugby club and he swam.

In the autumn of 1945, Barker returned to London to study at St George's Hospital and Medical School, on Hyde Park Corner. The buildings were dilapidated and tired after the war. Barker found the corridors dark and airless, and oddly quiet, given the noise of the traffic going past outside. The library had been hit by a bomb. Students and caretakers who worked late and alone sometimes reported sudden feelings of cold and depression, unexplained footsteps and presences nearby. Barker, who was twenty-one, began to collect their experiences. He spoke to nurses at the hospital, which was built in the 1830s, about an apparition that would appear next to the beds of severely ill patients in Williams and Marie Tempest wards and was said to presage their deaths. There was another ghost (who was perhaps the same), of a young nurse who had fallen to her death down a steep staircase after having an affair with a patient in 1926. A night nurse told Barker about glancing up to see a ward sister sitting at a desk under a light, only to look again and find that there was no ward sister on duty. Others complained of being gripped by paralysis, unable to move from their chairs when they heard footsteps approaching, which eased when the footsteps went away.

One evening, outside the school, Barker met a fellow student, 'C.P.', coming out of the doors, pale and terrified. According to Barker, C.P. was normally 'jovial, apparently phlegmatic, and definitely very sceptical of all psychical phenomena'. They went to a nearby pub, where, over a couple of drinks, C.P. told Barker that he had been alone in the school library, standing at some shelves, when he found himself shivering and sensed a presence behind him, looking over his shoulder. C.P. heard a scraping noise, like the sound of a metal ruler being pulled over an iron grating next to him, and ran out of the building. 'He rarely spoke of the event afterwards and did not want to be questioned about it,' Barker wrote.

A few months later, Barker stayed up late in the library with another friend, to see what would happen. The disturbances among the shelves were usually blamed on the ghost of John Hunter, a charismatic surgeon and tutor at the school in the late eighteenth century, who collapsed and died at the hospital when he was sixty-five. Hunter suffered from what one pupil described as 'a very irregular spasmodic affection at his heart', which would flare up whenever he was distressed or overworked. Hunter sensed that a sudden agitation might kill him one day. 'My life is at the mercy of any scoundrel who chooses to put me in a passion,' he said. In the autumn of 1793, there was a dispute among the surgeons at St George's and Hunter was asked to mediate. On the morning of 16 October he rose, inspected the nightly delivery of disinterred corpses to his rooms, prepared his dissecting tools, breakfasted well and went out to visit patients.

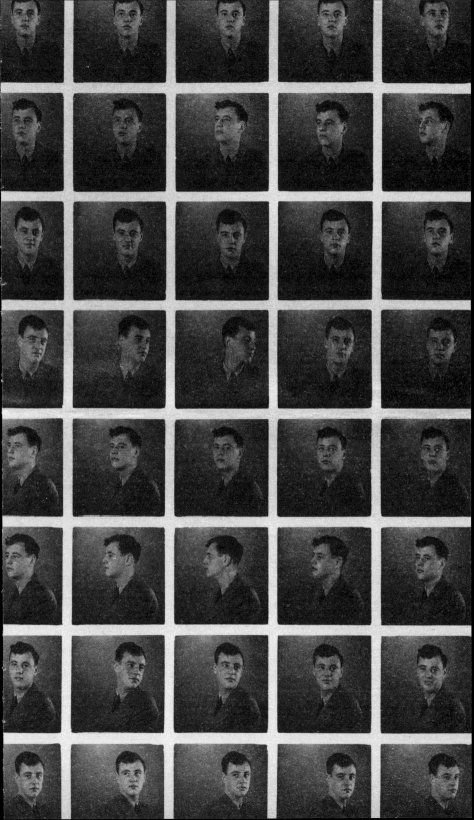

During the hospital board meeting, in the afternoon, Hunter flew into a rage. 'He had several words with the Surgeons, which brought on his complaint,' his pupil wrote. Hunter collapsed and died, felled by his passion. In the library, Barker tried to rouse his ghost by means of a planchette – a small wooden board used for automatic writing – and asking questions which Hunter could answer through 'spirit raps', sharp sounds emanating from the walls. Barker and his friend sat quietly, trying to decipher one knock for yes and two knocks for no from the intermittent tapping of the library's old heating pipes. 'We experienced no physical manifestations as such, although we nearly scared ourselves to death,' he recalled.

Towards the end of their studies, Barker and Jane became engaged. Late in 1947, they found an empty sitting room at the school, near the women's cloakroom, to go for a 'cosy chat', as Barker described it. They were alone and the door was slightly ajar. Half an hour later, at around 11 p.m., the door suddenly flew open, 'as if it had been blown by a powerful but non-existent wind,' Barker wrote. The room rapidly grew cold and he and Jane became very afraid.

We decided to leave immediately and rushed out of the door into the passage, leaving the lights on, my fiancée going first. Just as we left the room, we both heard a frightful crash behind us, rather like the sound of a large oxygen cylinder toppling on to its side, coming from inside the room in which we had been sitting.

They ran out into the street. Then Barker went back inside, on his own, to investigate. He groped along the passageway and back to the sitting room, expecting to find a cause for the noise, or some sign of disorder, but there was none. 'All was just as we had left it, and perfectly quiet,' he wrote almost twenty years later, in a collection of reminiscences for the school magazine. 'What does it all mean? Who or what were these restless agencies striving to attract our attention? What did they want?'

*

One Saturday in August 1955, Barker was on duty at St Ebba's, a sprawling, one-thousand-bed mental hospital outside Epsom in Surrey, when he was asked to examine a young man who had just arrived. The patient was in his early twenties, with dark hair and sharp features. 'He had an immature face and a tensed expression,' Barker wrote. The man, who described himself as a lorry driver from High Wycombe, was five foot eight inches tall, and slender. He glanced around the room as if it might contain something dangerous.

Barker was thirty-one and midway through his psychiatry diploma. St Ebba's was part of the Epsom Cluster, a group of five large hospitals built in the Surrey countryside at the turn of the century to receive mental patients from all over London. After the First World War it had been used to treat shellshocked soldiers. The patient had been transferred to St Ebba's from the psychiatric unit of Middlesex Hospital with few notes and no diagnosis. When Barker asked him to lie on a bed for a physical exam, the man became irritable and aggressive, saying it was a waste of time.

After a few minutes of resistance, the patient lifted his shirt to show Barker his torso, which was criss-crossed with surgical incisions. 'I was astounded at the appearance of his abdomen which seemed to be a solid mass of scar tissue,' Barker noted. On the patient's back, Barker found traces of twenty lumbar punctures. He listened sympathetically as the young man then related a vague and convoluted medical history. 'I really imagined that some of his answers made sense to me at the time,' Barker recalled. But he was also puzzled that the man, who seemed physically well, had required so many operations. Barker left the room to consult a senior colleague, who told him that this could be a case of Munchausen's syndrome.

Barker admitted the patient, whom he later called 'Maurice', sedated him and put him in one of the hospital's locked wards for the night, fearing that he might try to run away. On the Sunday morning, Barker returned to St Ebba's to find that Maurice had destroyed furniture on the ward, insulted the nurses and kept the other patients awake through the night, shouting. He had Maurice locked in a smaller room but the patient promptly escaped through a narrow side window. Later that day, Maurice's father, who was in fact a lorry driver, arrived at St Ebba's and told Barker that his son had been admitted to more than a hundred hospitals in the last few years. Within days, Barker heard that Maurice had been treated at a hospital in Chelsea and had run away again. 'I then began to devote much thought to this nervous and hate-filled young man,' Barker wrote. He was curious about Munchausen's syndrome, the unusual condition which seemed to have overtaken Maurice, and humiliated by his own inadequate encounter with it. In 1956, Barker qualified as a psychiatrist. He spent the next four years seeking out nine other Munchausen's patients in hospitals around the country for his doctoral thesis, which was one of the first clinical studies of the condition.

Maurice remained the principal character in Barker's thesis. In April 1957, with the help of Maurice's father, Barker traced him to a hospital outside Aylesbury. Barker arrived with a colleague, unannounced, and asked Maurice if he would undergo some psychological tests. The meeting was uneasy. Maurice recognised Barker, although he was not sure from where. There were times when he became aggressive and Barker had to leave the room. The rest of the time, the psychiatrist watched and listened to Maurice, wondering how 'such an insignificant little man' was capable of deceiving skilled and experienced doctors. 'I also wondered what made him behave like this, and why he had to have so many operations,' Barker wrote. 'At times I felt I could see why it was quite plainly, but then, in a flash, the inspiration would desert me.' For a time, Maurice found work in a hospital mortuary and seemed to settle. But two years later, a friend of Barker's called him from a casualty department in Wandsworth. Maurice had walked in, asking for a head X-ray after claiming to have fallen out of a bus on his way back from London airport. Barker arranged to have Maurice transferred to his care at Banstead Hospital, in Surrey, where he was part of the psychiatric team. He wrote a warning for the hospital's nursing staff on Maurice's notes: 'Not a single statement of his should be believed.'

For the next two weeks, the two men saw each other almost every day. Barker would arrive at work to find Maurice sitting in his office. The young man's mood would alter several times within a single conversation. Maurice, who was twenty-five by this time, could be ingratiating and ask for favours, and then provoke Barker and try and cause a row. One day, Barker softened and gave Maurice an afternoon pass; he came back at 1 a.m., drunk, with his girlfriend. Soon afterwards, he broke out of a bathroom window, climbed down a thirty-foot drainpipe and absconded to London, where he sought treatment for what he claimed was a fractured skull. He was brought back to Banstead by the police.

'At about this time, I wondered whether a pre-frontal leucotomy would help him,' Barker recalled. A leucotomy, also known as a lobotomy, involved the severing of the connections to the frontal lobes of a patient's brain. The surgery was a crude and radical step. Lobotomies had fallen out of favour, sharply, during the fifties because of the brutality of the procedure and the availability of new, much more reliable, pharmaceutical treatments. There was also no evidence to support the intervention in the limited literature about Munchausen's. Nevertheless, Barker theorised that the operation might work in Maurice's case 'to reduce his drive' and constant pestering of hospital staff. Maurice's parents, whom Barker described as a pleasant working-class couple, gave their consent. Maurice was frankly excited. 'Hurry up and get on with it,' he told his psychiatrist. 'It's what I have wanted for the past ten years.' He was thrilled while having his head shaved. Barker noted that this may have been 'a manifestation of his death instinct (Thanatos)'.

On 28 April 1959, holes were drilled in the sides of Maurice's skull and a leucotome – a silver surgical instrument with an extendable blade – was put inside. When a plunger was depressed, the blade opened and the leucotome rotated, cutting away at the tissue towards the front of Maurice's brain. For a month, Barker believed he had cured his patient. In triumph, he wrote to Richard Asher, the endocrinologist who had named Munchausen's syndrome in 1951. Asher did not reply. Then Maurice began to deteriorate. He drifted in and out of Banstead, going missing for days at a time. One night he returned, 'affectively flattened'; he 'was without insight, and had no apparent sense of right or wrong', another doctor observed. On 24 June, less than two months after he was lobotomised, Maurice turned up at a hospital in Romford in east London, complaining of headaches and weakness in his arms. He claimed he had been in a motorcycle crash. Three weeks later, he was in prison in Oxford after stealing a car.

Barker's treatment of Maurice stayed with him. In 1962, he wrote a letter to *The Lancet* opposing the use of leuco-tomies for the chronically mentally ill. In his doctoral thesis, Barker wrote a searching, introspective twenty-eight-page account of his treatment of Maurice and wrestled with the perilous role of the medical professional when dealing with Munchausen's syndrome – how a well-meaning doctor could deepen, and even realise, a patient's fantasies. 'The attitude of the surgeon is also important,' Barker wrote, 'for he may be unwittingly drawn into the patient's scheme of things.' Barker cited the analysis of Karl Menninger, in Kansas, who had studied Munchausen's under the label of 'polysurgery' and 'polysurgical addiction' in the 1930s. Menninger observed that a doctor's desire to heal and understand a patient's condition was a powerful, catalysing element in the encounter. 'Unconscious motives combine with conscious purposes to determine the surgeon's election to operate, no less than the patient's election to submit to the operation,' Menninger wrote. Barker also quoted Edward Weiss and Oliver English, two psychiatrists in Philadelphia who were pioneers of psychosomatic illnesses, who warned that the combination of a self-harming Munchausen's suf-ferer and a doctor eager to operate could end in the 'near evisceration' of the patient.

Doctors and scientists are not immune from delusions, or from being caught up in other people's. We like to imagine that they work on a higher plane of reason or operate with greater doubt. But there is an argument that they are more

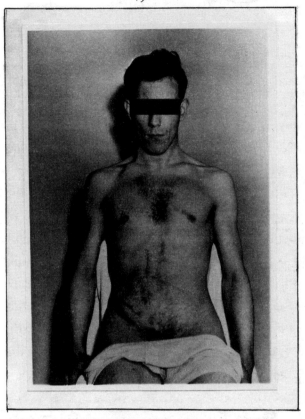

Case 1.   M.H.   ("Maurice").

Aged 25

"Mixed" abdominal and neurological type
of Munchausen Syndrome.

(Note extensive abdominal scars - "Railroad abdomen")

susceptible than the rest of us. The best researchers work on problems where patterns are hidden and the stakes are high. They crave a novel explanation of the world. In 1988, Brendan Maher, a psychology professor at Harvard, compared the making of scientific theories to psychosis. 'The necessity for a theory arises whenever nature presents us with a puzzle,' he wrote. 'Puzzles demand an explanation.'

The difference between science and madness is correcting your explanation when it doesn't map on to the world. After he failed to cure Maurice, Barker renounced the use of leucotomy but he did not seem to question his logic or the intense, involved approach that had led him to recommend the surgery. The data never supported the idea. If anything, it did the opposite (another Munchausen patient that Barker studied had also undergone a leucotomy, with no improvement). The only person who had really embraced the intervention was Maurice, of whom Barker had said: 'Not a single statement of his should be believed.' It was part of Barker's humanity as a doctor that he was willing to listen deeply to his patients and to try and perceive the world as they did. It was part of his vulnerability as a researcher that he believed there was a solution that only he could see. Barker submitted his doctoral thesis in January 1960, eight months after Maurice's lobotomy. He noted that Maurice's father had recently visited him, urging him to resume treatment of his son when he was released from prison. 'He has complete confidence in you,' Barker quoted Maurice's father as saying. The psychiatrist steeled himself for another attempt. 'Perhaps I shall have to try all over again.'

★

The work took its toll. In the late fifties, when he was looking after Maurice, Barker became obese. Weight massed on his heavy frame. A double chin hung over his tie. Barker ate fried breakfasts and smoked a pipe, like his father had. The long days and hospital corridors confined him. After his breakdown in Dorset, Barker was diagnosed with high blood pressure and was refused life insurance, on account of his health and a history of prostate conditions in his family. By the time he reached Shelton, he had been working in mental hospitals for eight years. The fantastical and the rare were obscured, day after day, by the mundane and the hopeless. One of the first things that visitors saw, after entering the hospital's rather grand administration block, were silent patients polishing the dark linoleum floor to an everlasting shine.

Barker coped with the futility, in part, by publishing ideas and research on an eclectic array of subjects. In 1958, while he was at Banstead, he surveyed forty-two mental hospitals on their use of ECT, asking which anaesthetics they used and what their side effects could be. He calculated a fatality rate of 0.0036 per cent associated with the treatment. For another project, Barker explored bone marrow problems among epileptic patients, who received medication for their convulsions. At Herrison, he became interested in altering patients' environments: integrating wards for long-stay and acute patients, and encouraging the sexes to mingle. At Shelton, he co-wrote a paper describing 'shopping days', when a local department store set up a temporary display, complete with fitting rooms, in the main hall. In successive months in early 1964, Barker appeared in the letters pages of *The Lancet* arguing for the establishment of a new official body for psychiatrists (which was set up later that year) and for the reform of the way that mental hospitals administered home visits by doctors.

He worked with an intensity born from a belief that almost any interaction with a patient might lead to a new paper or a notable result. In 1964, Barker learned that one of his former Munchausen's case studies, an orphan and former prostitute who swallowed open safety pins, was seeking treatment again in London. He arranged to have her admitted to Shelton, 150 miles away. After two months on a ward, Barker found the woman, 'Mrs BM', a job as a cleaner in a care home for the blind, from which she ran away. Barker circulated a description of Mrs BM to seventy hospitals across the Midlands until he tracked her down in Birmingham and brought her back to Shelton. The following year, he organised a clinical workshop with a colleague, Dr Sophia Lucas, who specialised in hypnosis, and conducted a session with Mrs BM in front of an audience of visiting psychiatrists, an account of which was published in the *American Journal of Medical Hypnosis*.

In Shropshire, though, there were signs that Barker took

better care of himself. He hung his pipes, along with old pub tankards, in a line along the hall of Barnfield. He played with his children. He became an avid collector of clocks. At any given hour, three grandfather clocks and a cuckoo clock would chime throughout the house. To get some exercise, Barker took up surfing. At a time when the sport had barely arrived in Britain, the psychiatrist bought a red-and-white surfboard and roped it to the top of his Ford Zephyr. When the family took their summer holidays in Woolacombe, in Devon, Barker would either disappear for the day to Newquay for some proper waves or his children would watch him – a bulky figure in a wetsuit, twenty years older than most of the other longboard pioneers – trying to catch a ride in the shallows. At the end of the day, Barker would come back to the hotel with cuts on his legs.

As his career progressed and his family grew, there was little or no space in Barker's life for the occult but he made time for it anyway. On quiet weekends, Barker took his eldest son, Nigel, on trips to haunted houses. Nigel was six or seven years old at the time. He preferred to stay in the car. When Barker worked at home, his children used to run up and down the corridor outside his study until he came out and told them to cut it out. They called him 'Dadah'. When their father's study door opened, the children got a glimpse of the crystal ball that he kept on his desk.

*

In the summer of 1965, two years after he arrived at Shelton, Barker read a letter in the *British Medical Journal* about the death of a forty-three-year-old woman in Labrador, in Canada. The patient, Mrs AB, was the wife of a fur trapper and the mother of five children, who lived on the banks of the Nauskapi River, in a trading outpost called North West River. The outpost was cut off for weeks at a time during the winter but it had a small, well-equipped hospital, the Emily Chamberlain, which was painted green and over-looked the water, and a team of doctors that went out on dog sleds, boats and small aircraft to treat nomadic Innu com-munities, which were strung for hundreds of miles across the interior. A few times a year, a surgeon would fly to the outpost to carry out minor surgeries.

Mrs AB had complained of incontinence but she was other-wise tough and healthy. The wives of fur trappers in North West River lived without their husbands for months at a time. They split wood, shot partridges, made clothes, which they sold at the trading store, and fished through holes in the ice. In March 1965 Mrs AB was admitted to the hospital to undergo a repair to her vaginal wall. She was nervous but the surgery seemed to go well. The operation lasted less than an hour. 'It was all completely normal,' Peter Steele, a young British doctor who looked after Mrs AB, recalled. Shortly after she regained consciousness, however, the woman complained of pain on her left side and went into shock. Her blood pressure collapsed and she died. A post-mortem revealed that Mrs AB had suffered an adrenal haemorrhage – a rare failure of the adrenal glands – but had no underlying illness. The hospital team were baffled. 'It was shattering,' Steele said. 'It was as if she got up and died.'

In the days afterwards, the doctors learned that, as a child, Mrs AB had been told by a fortune teller that she would die at the age of forty-three. Her birthday had been the previous week and she had been sure that she would not survive the procedure. Steele and his colleagues shared details of the case in the *BMJ*:

On the evening before the operation she told her sister, who alone knew of the prophecy, that she did not expect to awake from the anaesthetic, and on the morning of the operation, the patient told a nurse she was sure she was going to die. These fears were not known to us at the time of the operation.

We would be grateful to hear from any reader who has had experience of a patient dying under similar circumstances. We wonder if the severe emotional tensions of this patient superimposed on the physiological stress of surgery had any bearing upon her death. – We are, etc.

Barker was intrigued. In 1952, when he had recently qualified as a doctor, he had treated a man in Gloucester who was also convinced that he was about to die. The man was in his early forties, and had been found wandering in the town. 'He was literally terrified – I have never seen anybody so afraid, so much so that we were unable to converse with him at all,' Barker later wrote. 'He would answer none of our questions, but kept shouting, "I'm going to die. I'm going to die. Please don't let me die."' Barker gave the man oxygen and aminophylline, a drug to ease his breathing, but the patient died about half an hour after being admitted to hospital and a post-mortem revealed no obvious reason why. Two years later, Barker believed that he hastened the death of a second man, who was complaining of chest pains, when he asked the patient if he thought he was about to die. 'A rapid and profound change came over him,' Barker wrote. 'He did not reply – in fact, he never spoke again.' The patient slumped back and died about a minute later, while Barker was listening to his chest. This time the post-mortem showed that the man had thickened arteries and weakened muscles in his heart, but again no conclusive cause of death.

In these cases, medicine seemed only partly able to explain what had happened. In 1942, Walter Cannon, the head of physiology at Harvard Medical School, had used the phrase 'voodoo death' to describe a potential biological mechanism by which someone could be frightened to death. Cannon proposed that a person could die as a result of an overload of their sympathetic nervous system and their adrenal glands. He confined his research to 'primitive people' and 'black magic', but Barker believed that the phenomenon existed in Western societies too. The possibility of dying of fear, or coming to some prophesied end, existed in the realm just beyond conventional science, which attracted him the most.

He made contact with the doctors in Labrador, asking for further details of Mrs AB, and joined their appeal for further cases in the medical literature. Through the autumn of 1965, the *BMJ*'s letters pages carried stories of patients seemingly fatally afraid, or who died at moments that had been fore-told: a twenty-one-year-old mother died six days after giving birth to her baby, as she had been warned that she would; an otherwise hale seventy-four-year-old man refused to believe his own recovery after a heart attack. He had made his will and written to his solicitor. He smiled when he was told that he was getting better. 'His condition improved rapidly, but he never doubted his end was near,' wrote a doctor from Barnet, in north London. 'Three weeks after admission, he suddenly collapsed and died.'

Some of the letters in the *BMJ* cited Cannon's work on voodoo death. An Australian psychiatrist mentioned the experiments of Curt Richter, a psychobiologist at Johns Hopkins Medical School in Baltimore, who had shown in the 1950s that rats would give up and die in a hopeless situation. In Richter's lab, the rats were placed in glass jars full of water. Unlike Cannon, who described an overwhelming state of fear and adrenaline, Richter witnessed a kind of resignation, in which the rats lost their will to survive; their heart rates slowed, their temperatures dropped, their breathing packed up. Often it didn't take much. Wild rats died after their whiskers were trimmed. Others expired moments after they were handled. 'The situation of these rats scarcely seems one demanding fight or flight – it is rather one of hopelessness,' Richter wrote. 'Whether they are restrained in the hand or confined in the swimming jar, the rats are in a situation against which they have no defence.' Richter concluded that the rats lived and died according to their 'emotional reactions', which could be positive as well as negative. Richter found that animals in the same situation could survive desperately – swimming for up to eighty-one hours in a jar, in one case – if they were lifted out at some stage and given a glimpse of escape. 'After the elimination of hopelessness, the rats do not die,' Richter wrote.

Barker's letters in the *BMJ* recounted his own experiences of patients dying from fear. But unlike the other doctors, who were mostly cautious and speculative in their tone, his letters were confident, almost argumentative. He aired the possibility that being frightened to death could be linked to precognition and the 'subliminal self', a concept that Barker compared to Freud's id, which existed outside time. Barker harangued fellow psychiatrists and scientists for being insufficiently open to the potential role of ESP and other explanations that lay outside mainstream medicine. 'An important but curious feature which emerges is the frivolous and irresponsible attitude of many people, including some men of science, towards this subject in general,' he wrote in the *BMJ* on 18 September 1965. 'What is now unfamiliar tends to be inadmissible and is therefore just not accepted, even despite overwhelming supportive evidence. Thus for generations the earth was

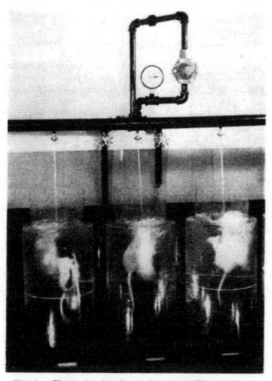

Fig. 1. Glass swimming jars, water jets, cold and hot water faucets, pressure gauge, and pressure regulator.

traditionally regarded as flat, and those who opposed this notion were bitterly attacked.'

The Labrador case ignited the research that led to *Scared to Death*, Barker's book on the subject, which in turn brought him to Aberfan. And it was his strident letters in the medical press that first caught the attention of Fairley, in 1965. The science writer had recently returned from a space reporting trip in the US and was combing through issues of the *BMJ* to see what he had missed. 'It was unusual for a well qualified doctor to be dabbling in such matters,' Fairley wrote. In one letter, Barker said that he wanted to know about how fortune tellers conveyed worrying predictions to their clients. Fairley wrote to Barker at Shelton and offered to host a meeting of astrologers, clairvoyants and card readers with the psychiatrist, if he could write it up for the *Evening Standard*.

*

The group met in a suite at the Charing Cross Hotel, above the railway station, not far from Trafalgar Square, in November 1965. Fairley had invited Katina Theodossiou, the *Standard*'s astrologer; William King, an Irish clairvoyant from Bristol; Tom Corbett, a crystal ball reader who practised in Chelsea; and Mir Bashir, a scholarly Kashmiri palmist who had arrived in Britain after the partition of India in 1947 and went straight to Scotland Yard to offer the police his services. Barker travelled from Shrewsbury with

Dr Lucas, his hypnotist colleague from Shelton, and a heavy tape recorder. Fairley had laid on a dinner of parma ham and melon, filet de bœuf and pineapple surprise. There was plenty of wine.

Before the group sat down to eat, Barker took each fortune teller in turn into a bedroom and interviewed them. Although he was an experienced researcher, there was a heedlessness to Barker's work on the occult. The psychiatrist did not observe formal boundaries or seek to protect himself in any way. It was the approach of a sceptic; only he was not a sceptic. As part of his research for *Scared to Death* Barker consulted a dozen psychics to see if they would predict the date and cause of his own death. Only one complied. But several of the fortune tellers impressed him with other observations. Four had observed that Barker was divided between his 'hobby' – which he regarded as his research into the supernatural – and his medical work. One recounted his breakdown and illness before going to Shelton. Another remarked that Barker himself might have a supernatural gift, something that he did not disagree with. A palm reader in London told Barker that he was marked by the psychic cross.

That night at the Charing Cross Hotel, he was matter-of-fact, almost brusque. 'I'd like you to be as brief as possible because there are a number of questions,' Barker said, opening his conversation with King, the clairvoyant. 'The first is have you ever predicted illness or even death in your clients at some future date?'

'I have predicted illnesses,' King said carefully. He was sixty-nine years old and had been a fortune teller since he was a boy. 'And I have predicted one or two deaths.'

'Only one or two?' Barker replied. He sounded unimpressed.

A few minutes later, Barker asked King if he thought other people could develop powers like his.

'Yes, if God gives it to them,' King replied, as if he had been asked the question many times before.

'Could I develop it for instance?' Barker asked.

'You are a psychic in yourself,' the clairvoyant replied immediately. 'But you don't know anything about it.' Then he hesitated. 'I'm sorry to say this . . .' King offered to stop if the psychiatrist wanted to pause the tape or change the subject. But Barker urged him to continue.

'No, it's interesting,' he said. His voice was softer.

'You are a psychic in yourself,' King repeated. 'And you should never be in the job that you're doing.'

Barker murmured in agreement.

'But you can do your job damn well,' King said.

'Thank you.'

'And you're respected. You're wanted in your job,' King continued. The clairvoyant steered the conversation to the safe patter of a fortune teller. 'I don't care if you're married and you have got six children. I couldn't care less,' he said. 'At times, you're alone. Very much alone.'

'Yes,' Barker said.

'And you don't know why,' King said. 'It is the ether. It is the outside complications of life.'

Barker tried to bring King back to his own potential abilities.

'But could I develop any powers like you have?' he asked again.

'Not now,' the clairvoyant replied. 'You're too late.'

The party broke up at two o'clock in the morning. At the end, Fairley took Barker aside and interviewed him for the pair of feature articles that ran in the *Evening Standard* the following week. It was eleven months before the Aberfan disaster but Barker already saw the question of people being scared to death as part of a broader inquiry into whether some events could be seen before they occurred.

When he was trying to explain premonitions, Barker often referred to *Foreknowledge*, a monograph written by Herbert Saltmarsh, of the Society for Psychical Research, in 1938. The first half of the twentieth century was awash with novel notions of time. Breakthroughs in particle physics, even if they were hazily understood, helped to undermine the idea of an orderly flow from one moment to the next, and to invigorate older and more mysterious theories of causation. Saltmarsh was a shipping agent in the City of London who had retired early because of ill health. In *Foreknowledge*, he examined the 349 cases of apparent precognition in the archives of the SPR and searched for a perfect example. 'Needless to say, I have not yet found one,' he wrote. Saltmarsh's monograph classified premonitions into various categories and set out some of the current theories about how they might occur. Speaking to Fairley that night, Barker cited Saltmarsh when he drew a distinction between our conscious experience of time, with its strict demarcations between the past, present and future, and a more porous experience available to our unconscious. 'I personally think that we should clarify our whole idea about time,' he said.

Versions of these ideas were quite common. Barker quoted the idea of 'the specious present', which was popularised by William James, one of the founders of modern psychology, in the late nineteenth century, to describe how our minds assemble a sense of things happening at any given instant. He explained to Fairley that, in his view, our conscious mind might have one specious present, which extends a few moments from now, while our unconscious may have a present which stretches further into the future. 'Suppose my normal specious present extends from noon until one second after noon, that's in my consciousness,' Barker said. 'And then my subliminal specious present extends from noon as far as one o'clock. All events happening up to 1 p.m. then are present events for my subliminal mind, and might be known to it.' Barker wondered what death would feel like if part of your mind had already glimpsed it.

*

When the Premonitions Bureau opened, it was not the first attempt to capture the visions of the British public. In the late 1920s, J. W. Dunne, an aircraft designer, wrote a popular book called *An Experiment with Time* which combined an account of his own precognitive dreams with a discussion of relativity theory and quantum physics. In 1902 Dunne was a young soldier serving in the Boer War when he dreamed of a volcano about to explode on a French colonial island, which would kill four thousand people. A few weeks later, he got hold of a *Daily Telegraph*, which reported the loss of some forty thousand lives after the eruption of Mont Pelée, on the Caribbean island of Martinique, and read about his dream in print. 'I was out by a nought,' Dunne reflected.

Premonitions, banal and tremendous, stalked him for years. Dunne's response was unsentimental. 'No one, I imagine, can derive any considerable pleasure from the supposition that he is a freak,' he wrote. By the end of the First World War, Dunne was consoled by advances in quantum mechanics that suggested the old order of time was collapsing: 'That, already, was in the melting-pot,' he wrote. 'Modern science had put it there – and was wondering what to do next.'

Dunne's own theory about how time worked, which he called serialism, was hard to follow, but *An Experiment with Time* was influential because it encouraged thousands of readers to keep dream diaries and to see if their presentiments materialised. Dunne emphasised that we should pay attention to trivial flashes of the future as well as things that seemed important. He liked to sit in the library of his club, pick up a novel, glance at the name of the protagonist and then jot down thoughts and images that came to him, to see if they predicted the plot. One day, Dunne picked up a book by

J. C. Snaith, a cricketer turned popular author, but nothing came to him except a peculiar image of a plain black, entirely straight umbrella, standing vertical – its handle resting on the pavement – outside the Piccadilly Hotel. The next day, Dunne found himself on a bus as it approached the hotel and noticed a figure walking along:

> It was an old lady, dressed in a freakish, very early-Victorian, black costume, poke bonnet and all. She carried an umbrella in which the handle was merely a plain, thin, unpolished extension of the main stick . . . She was using this umbrella – closed, of course – as a walking stick, grasping it pilgrim's staff fashion. But she had it *upside down*. She was holding it by the ferrule end, and was pounding along towards the hotel with *the handle on the pavement*.

While Dunne's work was popular in Britain, twentieth-century physics and psychology catalysed similar interest in prophetic dreams elsewhere in Europe. In 1933, a Jewish journalist in Berlin, Charlotte Beradt, began secretly writing down the dreams of German citizens soon after the Nazis came to power. Three days after Hitler was elected Chancellor, a factory owner dreamed that it took him half an hour of excruciating effort to raise his arm in salute during a visit by Joseph Goebbels. A thirty-year-old woman dreamed that all the street signs in her neighbourhood had been replaced by posters with a list of twenty words that it was now forbidden to say. The list started with 'Lord' and ended with 'I'. Later, the same woman dreamed that a squad of policemen hauled her out of a performance of *The Magic Flute* because a thought-reading machine – 'it was electric, a maze of wires' – had registered her associating Hitler with the word 'devil', when it was sung by Papageno and Monostatos during the opera. Beradt collected around three hundred dreams. Many involved bureaucratic absurdity – 'The Decree of the Seventeenth of this Month on the Abolition of Walls'; 'A Regulation Prohibiting Residual Bourgeois Tendencies' – which prefigured the totalitarian intentions of the regime. A Jewish lawyer dreamed that he was crossing Lapland to reach 'the last country on earth where Jews are still tolerated' but a smiling border official threw his passport into the snow. A green, safe land lay tantalisingly out of reach. It was still 1935.

Beradt posted her notes to friends or hid them in books, and published them after the war. In *The Third Reich of Dreams*, she wrote that these 'diaries of the night' seemed 'to record seismographically the slightest effects of political events on the psyche'. They were raw, untouched by hindsight and possibly prophetic for that reason. 'Dream imagery might thus help to describe the structure of a reality that was just on the verge of becoming a nightmare,' Beradt wrote.

In 1940, when Britain was threatened with invasion, the playwright J. B. Priestley delivered regular Sunday evening radio talks, called 'Postscript', which were heard by a third of the British population. Priestley was from Bradford. He evinced a patriotic longing from the scattered notes of bird-song or a day trip to the seaside. He was also a follower of Dunne; Priestley described himself as 'time-haunted'. In the early 1930s, he had travelled to the American West. Early one morning, he stood by a railing on the South Rim of the Grand Canyon, with the landscape shrouded in mist. Suddenly the mist lifted, the colours shone, and Priestley recognised the railing, the sky and the canyon from a vivid dream, years before. (In the dream, he had been sitting in a theatre when the curtain lifted and displayed precisely the same scene.) Priestley's plays, such as *Time and the Conways* and, later, *An Inspector Calls*, reflected his preoccupation with the order of time. He helped to publicise Jung's idea of synchronicity, which proposed that events could be linked by meaning rather than causation, in the English-speaking world.

In March 1963, a few months before Barker arrived at Shelton, Priestley appeared on the BBC arts programme *Monitor* to talk about time. Priestley was almost seventy years old and had become a beloved national figure. He equated a strict, materialist reading of how time passes – each second of our lives flowing remorselessly, one after the other, until death – with the intellectual barrenness of cap-italist consumption. 'The moment does not matter because it is only another little step towards final oblivion,' Priestley wrote in *Man and Time*, which was published the following year. 'It is all a tale told by an idiot.'

Priestley was struck by how earlier and non-Western cul-

120

tures were comfortable with more sophisticated notions of time. He proposed a model of three concurrent times (the present, the unconscious and a collective unconscious) which was a fusion of Jungian and psychical ideas, not unlike Barker's. Priestley compared living within the modern understanding of time to balancing on a rope that was fraying at both ends: scientists knew that time was unpredictable at both the planetary scale, because of relativity, and at the subatomic scale, because of quantum physics. So why should it flow steadily, ceaselessly, through human lives? Priestley described 'a world dominated by the worst idea of time men have ever had'.

Man and Time was part confessional, part manifesto. Priestley implored society to step off the 'inexorable conveyor-belt to nothingness'. During his BBC broadcast in 1963, the interviewer, Huw Weldon, invited viewers to send in their own unusual experiences of time. Priestley received around fifteen hundred letters, of which around a third appeared to come from followers of Dunne.

*

Barker wanted the bureau to be more than another collection of anecdotes. The Aberfan material had convinced him that it was no longer necessary to prove the existence of precognition. In an article for the *Medical News* in January 1967, two weeks into the experiment, Barker claimed that there were now more than ten thousand incidents recorded in parapsychology journals. 'We should instead set about trying to harness and utilise it with a view to preventing further disasters,' he wrote.

Like Beradt in Nazi Germany, Barker used the language of seismology to describe mental processes which might be operating at a deep level within the collective subconscious. He wanted an instrument that was sensitive enough to capture intimations that were otherwise impossible to detect. He envisaged the fully fledged Premonitions Bureau as a 'central clearing house to which the public could always write or telephone should they experience any premonitions, particularly those which they felt were related to future catastrophes'. Over time, the Premonitions Bureau would become a data bank for the nation's dreams and visions – 'mass premonitions', Barker called them – and issue alerts based on the visions it received:

Ideally the system would need to be linked with a computer, to help exclude trivial, misleading or false information . . . With practice, it should be possible to detect patterns or peaks which might even suggest the nature and possible date, time and place of a disaster so that an official *early warning* could then be issued.

'There might be numerous false alarms, particularly in the early stages, when the operators were inexperienced,' Barker conceded. He recognised that the bureau also faced a version of the quandary that haunted Jonah in the Old

Testament. God asked Jonah to prophesy the destruction of Nineveh. But Jonah reasoned that if the people of Nineveh believed his warning and repented, God would forgive them and Nineveh would not be destroyed after all. Jonah's prophecy would turn out to be false, and he would look like a fool. Befuddled and ashamed, Jonah ran away and ended up inside a whale.

If a calamity is averted, how can it generate a vision to precede it? 'Theoretically, there might be no premonitions since no disaster would have occurred,' Barker acknowledged. But it was worth a shot. There were plenty of cases of premonitions that appeared to have helped avoid certain disasters in the past. 'If only one major catastrophe could be shown to have been prevented by this means,' Barker wrote in his paper for the SPR later that year, 'the project would have more than justified itself, perhaps for all time.'

*

The bureau got its first major hit in the spring of 1967. At 6 a.m. on 21 March, the phone rang in the dining room at Barnfield. Barker came downstairs and answered. It was Alan Hencher, the Post Office switchboard operator, one of the Aberfan seers who, like Miss Middleton, claimed to experience physical sensations before a disaster.

'I was hoping not to have to ring you,' Hencher said. 'But now I feel I must.'

Hencher was coming off a night shift and was calling

to predict a plane crash. Barker made notes on a piece of Shelton hospital letterhead. Hencher was upset. He had a vision of a Caravelle, a French-built passenger jet, experiencing problems soon after take-off. 'It is coming over mountains. It is going to radio it is in trouble. Then it will cut out – nothing.' Hencher said there would be 123 or 124 people on board ('? say 124', Barker jotted down) and that only one person would survive, 'in a very poor condition'. Hencher couldn't tell where the crash was going to happen but he had had the feeling for the last two or three days. It was as if someone on the aircraft was trying to communicate with him. They were trying to make peace. 'While I am talking to you, I have a vision of Christ,' Hencher told Barker. He could see a pair of statues and was directed to the crash by a light flashing on and off. Barker's notes ran to the bottom of the page and into the corner. On the other side of the paper, he noted that he called Hencher back later for more details, but there were none.

It was an hour before dawn, on a Tuesday morning. Barker was already in an unsettled state. The previous day, he had been summoned to a meeting at the regional health board in Birmingham, where he had been reprimanded by Shelton's superintendent, Dr Littlejohn, for the publicity that his work was attracting. Barker's claim to have cured Mr X's infidelity with electric shock treatment the previous December had caught the imagination of Britain's tabloid press. The *People*, a Sunday newspaper, had managed to identify Barker's patient and his wife – a Mr and Mrs Candlin, in Shrewsbury – and

Shelton Hospital
Shrewsbury.

Mr Alan Hencher   6.0 am   21st March 1967

Aircraft Caravelle over mountain
will leave in Cog between Monday &
Sunday — it is coming over mountains. It
begins to show manage that it is in trouble.
Then it will cut out — time 123 — people
aboard. 123 people to going to crash, Surely
can't take off. I can't quite tell when
its going to happen. I Person saved — in
long a tunnel. ? where. Has had this

paid them £1,000 to share their story. Barker had refused to co-operate with the newspaper until reporters came to his house and he was advised by the regional health board to give an interview. But when a sensational description of Barker's treatment appeared, serialised over three consecutive Sundays in the *People* and featured on the ITV news in early March, Littlejohn was furious. 'Literally white with rage,' Barker recalled. He worried that the superintendent was trying to sack him.

At the meeting in Birmingham, Barker had been cleared of any wrongdoing. But he took the opportunity to warn Littlejohn and the local NHS board about the other research projects that he was pursuing. For the first time, the psychiatrist told his supervisors about both *Scared to Death* (Barker had recently delivered a draft of the book to his publishers) and the Premonitions Bureau.

'Littlejohn said nothing,' Barker noted, in a memo that he sent asking advice from a medical defence lawyer a few days later. The NHS officials, who had been sympathetic about the aversion therapy story, became alarmed. Barker was told that he would have to publish his book anonymously and remove his name from any association with the Premonitions Bureau, or risk losing his job.

'What shall I do? Am I to walk into a trap?' Barker asked the lawyer. He saw himself approaching an impossible situation, caught between his research, which transcended the conventional borders of psychiatry, and the suffocating strictures of Shelton. In 1963, Littlejohn had asked both Barker and Enoch to submit their research papers and journal correspondence to him before sending them for publication. The two young doctors had refused. 'Dr Littlejohn was not pleased and said he was left out of the picture,' Barker noted. He thought the older man was jealous.

Barker wasn't sure if he could alter his book or tell Fairley to stop collecting premonitions even if he wanted to. His account of his research to the lawyer had an edge of grandeur. He described momentous events sweeping him along, rather than esoteric research projects that he pursued in his spare time. 'I am one of those people who is interested in work and has had some success,' he wrote. According to Barker, his publisher had already decided that *Scared to Death* was going to be one of their 'big' books of 1968. 'The publicity could be considerable,' he wrote. Barker described his work on Aberfan as 'essential material and perhaps the largest study on precognition in existence'. The Premonitions Bureau, meanwhile, was entirely his idea and 'the logical outcome' of the Aberfan work. Calls and letters were coming in every day. 'At any time a major disaster could be forecast,' Barker wrote. He wasn't sure how much power Littlejohn and the hospital board really had. The meeting in Birmingham had lasted an hour. 'I was quite exhausted afterwards and nearly collapsed,' Barker wrote. He was reminded of his traumatic departure from Dorset.

After being woken by Hencher's telephone call that morning, Barker passed the prediction on to the *Evening Standard*. In the subsequent weeks, he made no effort to curb his extracurricular research or to stop drawing attention to himself. On 11 April, he and Fairley appeared on *Late Night Line-Up*, a chat show on BBC2, to publicise the bureau. Nine days later, a turboprop Britannia passenger aircraft carrying 130 people attempted to land in Nicosia, Cyprus, during a thunderstorm. The plane, which belonged to Globe Air, a new low-cost Swiss charter airline, was on its way from Bangkok to Basel, carrying mostly Swiss and German holidaymakers. It had refuelled in India and was on its way to its penultimate stop, in Cairo, when the pilots were advised the airport was closed because of heavy rain. The flight plan suggested Beirut as the back-up option but the captain, a British pilot named Michael Muller, decided to make an unscheduled landing in Cyprus, despite the bad weather.

By the time the plane reached the island, it had been in the air for almost ten hours. Muller and his co-pilot were almost three hours over their time limits at the controls. At 11.10 p.m., the aircraft was cleared to land at Nicosia, but came in a little high. Muller requested permission to make a circuit of the airport and try again. The control tower glimpsed the plane, its landing lights flashing through the low cloud, before it wheeled to the south and clipped a wing on the side of a hill – twenty-two feet from the summit – rolled over, broke into pieces and caught fire.

'124 DIE IN AIRLINER', the *Evening Standard* reported on its front page the following morning. (The final death toll was 126; two people who survived the initial impact were taken to a nearby UN field hospital, where they died.) At the time, the Nicosia crash was the sixth worst aviation accident in history. Fairley and Barker noticed the similarities with Hencher's prediction immediately. The *Evening Standard* published an account of Hencher's premonition alongside the news coverage that day. 'The Incredible Story of the Man Who Dreamed Disaster', the headline read. An accompanying photograph showed Archbishop Makarios, the island's Greek Cypriot president, picking through the wreckage.

*

Hencher was a gaunt forty-four-year-old man who lived with his parents in a council house in Dagenham, in Essex. The family had moved out of the East End of London before the war. Percy, Alan's father, had worked as a local government clerk. His mother, Rosina, stayed at home to look after the couple's three sons. The eldest, Eric, had served in the Commandos in Burma; the youngest, Ken, was a professional footballer for Millwall FC in the fifties before leaving the sport to become a customs and excise official. Alan, who had once been an apprentice to an optician, was the odd boy out. The Hencher family liked a drink; Alan preferred to read. He was affable but serious. He was proud of his collection of history books. In 1949, when he was twenty-six, he suffered a head injury in a car accident and was unconscious for four days. His precognitive ability began soon after. 'He was just different to the rest of them,' his niece, Lynne, recalled. 'He was very intense about everything.'

On the day of the plane crash, Fairley tried to call Hencher from the *Evening Standard* but failed to get through. Barker had arranged to speak to Hencher the following day. Shortly before one in the morning, the telephone in the dining room at Barnfield rang again. Barker came downstairs. It was the night-time switchboard operator at Shelton. Hencher had called the hospital, trying to reach Barker. He sounded agitated and the operator wanted to put him through.

Hencher came on the line and said that he was now worried about Barker's safety. He had been worried about him all day – that there might be some kind of accident. When Hencher thought of Barker, his mind filled with the colour black. He urged the psychiatrist to check his gas supply. But Barnfield didn't have a gas supply.

'Have you a dark car?' Hencher asked.

Barker's Zephyr was dark green.

'Be very careful,' Hencher warned. 'Look after yourself.'

Barker asked Hencher if he was telling him that his life was in danger.

'Yes,' the seer replied.

III

At 10 a.m. the next day, Barker dictated a four-page memo in his office at Shelton which he called 'Some Interesting Predictions and a Possible Death Sentence'. In the document, he outlined a medical history of Hencher and his apparently successful premonitions of the Aberfan disaster and the recent plane crash. Then the psychiatrist recounted Hencher's call during the night, and his response to being warned about his own fate:

My reactions to this were naturally to be somewhat
alarmed. I found it a little difficult to get off to sleep again
and have, of course, decided to take extra care while
driving. It would be wrong for me to say that I was not
frightened by a prediction of this nature. I intend keeping
a diary from now on and to record my reactions to this
on a daily basis. I suppose anybody who plays about with
precognition in this way to some extent sticks his neck
out and must accept what he gets. The important thing
though is for this information to be recorded so that if
anything does happen it should cause some interest and
may stimulate others to continue in this important work.
Of course, it is possible that this prediction as with the
others may not be fulfilled in a literal way. It would be

131

curious and remarkable indeed if Mr. Hencher could bring off a 'psychic "hat-trick"'. Having recently written a book on people who were 'Scared To Death,' I am perhaps beginning to feel what this would be like.

I found Barker's memo, along with some of his letters, in a brown envelope marked 'Prediction 3a' in the archives of the Society for Psychical Research, which are kept in the Cambridge University Library. The other letters in the envelope are all from the spring of 1967 and together they show Barker's complicated attitude towards the occult. He could be credulous, or laconic; doubtful, yet insinuating. He was interested in buying a house close to Shelton – a former pub which had served an old racecourse – in part because it was haunted. The house was called The Squirrel and a ghost named Joe was said to walk around upstairs.

'Personally, I would be fascinated to buy such a house from the scientific point of view,' Barker wrote to Guy Lambert, a former president of the SPR, in April. 'But I do not know whether this would apply to my wife and children.' Jane was eight months pregnant at the time. When Lambert replied to Barker's letter, suggesting possible natural causes for the strange sounds and disturbances at The Squirrel, the psychiatrist wrote back: 'I am not "sold" on E.S.P., but I have always maintained that a purely physical view is far too restricted and would be unlikely to explain everything.'

Two weeks later, however, buoyed by Hencher's apparent premonition of the Cyprus crash and unnerved by the warning about his own life, Barker sounded much more like a man feeling the surface of a never-seen object in the dark. 'Look carefully at Mr Hencher's prediction about me which thank God has not yet been fulfilled,' he wrote to Lambert on 8 May, after sharing a copy of his 'Death Sentence' memo. 'I do not know whether we are on the edge of something important. Personally I have my doubts, but one just does not know what is going to come up when one embarks on a scheme like this.'

*

'Prediction 3a' also contained a letter from Miss Middleton. When the bureau opened, Barker had circulated a request for premonitions to around a hundred potential percipients, including those who had contacted him for the Aberfan experiment. Miss Middleton's letter, which was from May 1967, was labelled 'Another "prediction reactor"' by Barker, to indicate that she belonged in the category that interested him the most. By that time, Miss Middleton had contacted the bureau several times. In mid-March, she had dreamed of her dead father sitting in her front room, taking a telephone call about danger at sea. She notified Barker. A few days later, the *Torrey Canyon*, an oil tanker on its way to Wales, ran aground between the Scilly Isles and Cornwall, causing Britain's worst oil spill to date. On 10 April, Miss Middleton wrote to Barker again, warning of a tornado or a hurricane on the west coast of the US. Eleven days later, there was an outbreak of more than forty tornadoes across five mid-western states, one of which killed thirty-three people in Oak Lawn, a suburb of Chicago. Barker congratulated Miss Middleton. 'This has certainly been a prophecy fulfilled,' the psychiatrist wrote, even though the damage was almost two thousand miles from America's western seaboard.

Later, Fairley recalled that he opposed engaging either way with people who sent their hunches to the Premonitions Bureau. Barker had no such compunction. He cajoled, he encouraged and he needled. Ten years after his troubling experience with Maurice, the Munchausen's sufferer, the psychiatrist again gave credence and attention to men and women whose illusions had not been previously taken seriously. His intellectual aspirations combined with their desire to be believed, and either he did not imagine the consequences or he wanted, at some level, to bring on those consequences, whatever they might be.

After Barker harried Hencher for further details of his plane crash prediction, Hencher complained to the psychiatrist that the research was having 'an adverse effect on his mind'. Barker reassured Hencher and urged him to carry on. He flattered Miss Middleton, who was delighted to be corresponding with such a distinguished doctor. 'I send my very best wishes for this experiment,' she wrote.

When Hencher predicted that Barker might die, the psychiatrist did not ignore or play down the suggestion, or consider pausing what he was doing, but used it instead as material to further his investigation. He sent a copy of the

'Death Sentence' memo to Miss Middleton and asked her if she was worried too. 'Re your own personal safety, about the time Mr Hencher was thinking about you, I had some personal concerns,' she replied. 'I remember thinking that the work you were doing was so important a prayer must be said for your well being.'

On 23 April 1967, Miss Middleton sent in a vision of an astronaut on his way to the moon. 'This venture will end in tragedy,' she wrote. The spaceman that she saw was 'petrified, terrified and just frightened'. Miss Middleton enclosed a drawing with her premonition, as she sometimes did, of an astronaut crouched inside a crude, spherical craft.

She posted her message in Edmonton, at 5.30 p.m. on a Sunday. At around the same time, Vladimir Mikhailovich Komarov, a forty-year-old Soviet cosmonaut, was napping in the living compartment of the Soyuz 1 spacecraft, on his twelfth orbit of the Earth. Komarov had blasted off from the Baikonur cosmodrome, in Kazakhstan, shortly before dawn that morning. It was the first piloted space mission launched by the USSR for more than two years. News of the launch had been released by TASS, the Soviet news agency; Radio Sweden had broadcast the information at 7 a.m., so it is possible that Miss Middleton knew there was a man in space. But few details were known.

The first Soyuz mission was technically hazardous and conducted under significant political pressure. The idea was for Komarov's craft to be joined by a second Soyuz spaceship in orbit the following day, for two cosmonauts to transfer from one vehicle to the other and for both modules to return to Earth. Nothing like this had been attempted before. The goal was partly to surprise and unsettle the Americans, who were now thought to have surpassed the USSR's capabilities in human space flight. But the preparations were not good. Three automated tests to rehearse the mission failed; three prototype spacecraft were destroyed in the process. On 14 April, nine days before the launch, engineers identified 101 things wrong with Soyuz 1 and Soyuz 2. The USSR's talismanic chief rocket engineer, Sergei Korolev, had died a year earlier and there was a feeling of unease. 'There is no confidence,' Lieutenant-General Nikolai Kamanin, who led the cosmonaut training programme, wrote in his diary. Nonetheless, the mission went ahead. It was timed to take place just before the USSR's annual May day celebrations. Before blast-off, Komarov dedicated his flight to the fiftieth anniversary of the Bolshevik Revolution.

Eighteen minutes into the mission, Komarov's problems began. A solar panel which provided internal power to the spacecraft failed to deploy. Part of the telemetry system wasn't working either and a 'sun-star' sensor, which was supposed to help position the Soyuz through its perilous re-entry into Earth's atmosphere, had fogged up. The fuel pressure dropped and the temperature inside the capsule began to fall. Six and a half hours into the flight, it was clear that the historic docking with Soyuz 2 could not take place. The mission was curtailed and the second Soyuz launch was abandoned. Komarov was told to rest and prepare to return

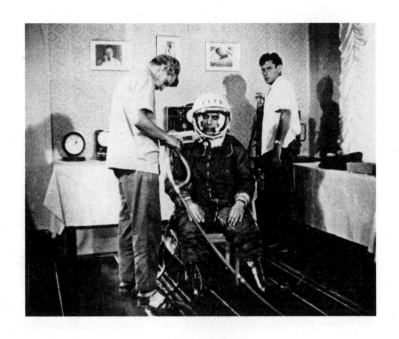

to Earth. He would land on his seventeenth orbit. Komarov
was a distinguished, experienced pilot. This was his second
time in space. His father had been a janitor. Even though
the flight was badly off track, he didn't let it show. 'I feel
excellent,' he said. 'The mood is good.'

Komarov was the oldest cosmonaut in the Soviet space
programme. He was close to Yuri Gagarin. Gagarin had
been there when the hatch at the top of Soyuz 1 was sealed.
For the remaining hours of the flight, Gagarin's was the
main voice that Komarov heard as he attempted to steer the
spacecraft largely by hand and eye, swinging from the dark
side of the Earth into the light, instruments failing around
him, through the narrow angle of descent that would take

him home. Gagarin radioed 'the kindest wishes for a soft, good landing'. Komarov replied: 'Thanks for your wishes. Not much time before we meet, so I'll see you soon.'

On his first two attempts at re-entry, Komarov's engines failed to fire and the spacecraft bounced back into orbit. On the third occasion, with the Soyuz's batteries running low, and flying almost entirely by his own reckoning of the stars and the planet beneath him, Komarov made it through. 'Thank you to everyone,' he said. He moved to the craft's middle seat and prepared for landing. He breathed a sigh of relief. Then the spacecraft's parachutes failed. The Soyuz fell until it hit open ground in southern Russia, not far from the Kazakh border, at 6.24 a.m. The wreckage caught fire. Komarov was burned to a brick. His remains, when they were recovered, measured roughly eighty centimetres by thirty. Rescue teams put out the burning Soyuz by covering it with dirt.

For several hours, senior Soviet officials did not know what had happened to Komarov. There were snatches of information. The spacecraft had broadcast an 'Emergency-2' signal as it careened through the atmosphere. Knowing the truth, the rescue teams at the crash site turned off their radios. In Moscow, Komarov's wife Valentina and their two young children, Evgenii and Irina, waited. It was a cloudy day. Valentina had only found out that Komarov was in space twenty-five minutes after he blasted off. 'My husband never tells me when he goes on a business trip,' she had joked with reporters.

The clouds in Moscow turned to rain. The telephones in the house stopped working. In doubt, we look for signs.

We see auguries. We extrapolate freely. The wife of a fellow cosmonaut arrived, unannounced, to sit with Valentina. Irina noticed that at this point her mother began to shake. A black Volga limousine pulled up outside the house and a general approached the door.

An investigation into the crash found that the Soyuz's parachutes had a design flaw. They were never going to work. Komarov's difficulties during the mission had very little to do with his demise. He had overcome them with great tenacity and sangfroid and then been killed by something that was always going to go wrong. He was never coming back. By happenstance, the unrelated misfirings on Komarov's rocket did save the lives of the three cosmonauts who were due to meet him in orbit the following day, because the parachutes on Soyuz 2 had been designed the same way. If it had gone into space, they would have failed too. Komarov was the first person to die during a space flight. His ashes were interred in the walls of the Kremlin.

Barker was thrilled by Miss Middleton's premonition. 'You were spot on,' he wrote. 'Well done!'

*

In the 1690s, a young tutor named Martin Martin was commissioned to map and document life in the western islands of Scotland. Martin, whose Gaelic name was Màrtainn MacGilleMhàrtainn, grew up on a small homestead at the northern end of Skye. He devoted thirty-five pages

139

of his study, which was published in 1703, to second sight. Islanders saw friends fall from horses when they were far away from home. Phantom wedding processions moved through the fields. Children saw bodies laid out on sideboards, or dead relatives who walked among them. A man in Flodgery dropped his knife in the middle of supper when a corpse materialised among the plates. Several times, Martin was seen a hundred miles from where he was. Villagers rode out to meet groups of horsemen or funeral processions that they saw crossing the hill, only to find that they were a day or a week ahead of time.

The Gaelic phrase for it was 'two sights', or *an da shealladh*. There was a visionary as well as an actual perception of the world. Patterns recurred in second sight. A spark from a fire falling on your arm meant that you would cradle a dead child. A woman who appeared on a man's left would become his wife. If a chair appeared empty, even though someone was sitting in it, they were soon to depart this world. If you heard voices or saw trees in an empty, barren place, there would probably be a house built or an orchard growing there soon.

Other visions were eerily particular. A man on Lewis was haunted by a doppelgänger who harangued him when he was working on the land, but remained quiet and polite when inside his house. The man grew tired of the vision and threw a piece of burning coal at it. In return, the spirit beat him black and blue. The minister was summoned and the congregation gathered around the haunted man to pray. It was no use. A boy from Knockow kept seeing a coffin over his shoulder, until he volunteered to be a pall bearer and the presentiment passed. A woman on Skye was tantalised by the vision of a figure who appeared dressed just like her but always had her back turned. 'Until the woman tried an experiment to satisfy her curiosity,' Martin wrote. The woman put her clothes on back to front, to coax the spirit to turn around. 'And it fell out accordingly,' Martin explained. 'For the vision soon after presented itself with its face and dress looking towards the woman, and it proved to resemble herself in all points, and she died in a little time after.'

Martin anticipated the scepticism of his readers. 'The second sight is not a late discovery seen by one or two in a corner, or a remote isle,' he insisted. It was experienced by both men and women; it was not obviously inheritable; it did not manifest with drink. It was not something that people especially enjoyed. Martin compared seeing the future to the contagiousness of yawning or the power of magnetism – other observable phenomena that men of science were yet to properly explain. 'If we know so little of natural causes, how much less can we pretend to things that are supernatural?' he asked. Martin argued that the islanders were not simply impressionable. They did not believe every single portent would come true. 'But when it actually comes to pass after-wards it is not in their power to deny it without offering violence to their senses and reason,' he wrote. Most people accepted premonitions as part of the mysterious workings of time. Others sought to ward them off, which is not the same as disbelieving. Martin knew of a man named John Morison, on Lewis, who had sewn a medicinal plant into the neck of his jacket to keep the visions away.

A Description of the Western Islands of Scotland stayed in print throughout the eighteenth century. It was a major source for Samuel Johnson's exploration of the Hebrides seventy years later and his own investigation of second sight. 'This receptive faculty, for power it cannot be called, is neither voluntary nor constant,' Johnson noted. 'The appearances have no dependence upon choice: they cannot be summoned, detained, or recalled. The impression is sudden, and the effect often painful.' Like Martin, Johnson found that seers seemed to derive no benefit or pleasure from their ability. 'It is an involuntary affection,' he wrote. And like Martin, Johnson reported that *an da shealladh* was nowhere near as widespread as it used to be. Prophecy is almost always in decline.

Martin's study was commissioned by the Royal Society, which had been founded in 1660 to disseminate factual knowledge of the world. The society's motto was '*Nullius in verba*': to take nobody's word for it. Martin did make one objective observation about the second sight of the Western Islanders, which was that it appeared to be a social phenomenon, and that when people moved away from the Hebrides, the future no longer appeared to them:

Four men from the isles of Skye and Harris, having gone to Barbadoes, stayed there for fourteen years; and though

142

they were wont to see the second-sight in their native country, they never saw it in Barbadoes; but upon their return to England, the first night after their landing they saw the second-sight, as was told to me by several of their acquaintance.

We see the world as our community sees it. We are drawn into each other's scheme of things. In the twentieth century, neuroscientists observed how some of our priors can be baked in by the places and the circumstances that we find ourselves in. In 1947, psychologists at Harvard found that ten-year-olds from a poor neighbourhood in Boston were more likely to imagine coins to be much larger than they really were, compared to children from a more prosperous part of the city. Deep-sea fishermen have more superstitious rituals than those who fish closer to shore.

Our propensity to sense objects that are not there – a tendency described as 'perceptual hunger' by Jerzy Konorski, a Polish student of Ivan Pavlov, in 1967 – increases as we feel less and less in control of what is happening to us. In the 1970s, researchers at the Institute of Higher Nervous Activity and Neurophysiology in Moscow showed slides to five men at various stages of a parachute jump. Some of the slides contained numbers embedded in jumbled patterns, others were pure noise – random black dots. The parachutists' ability to see the underlying numbers peaked as their aircraft was taking off, when they were excited and alert. But as the parachutists' stress increased, their perceptions

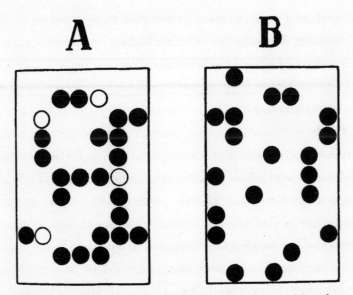

Fig. 1. Examples of visual images. A—a figure with noise (open circles in the outline mark the effect of distortion on the symbol), B—"pure noise."

started to go awry. There were more false alarms. At the moment they were about to jump out of the plane, they were most likely to see things that were not there. Life was hard in the Hebrides. Death came often and for no reason. Discerning logic or causation where there is none is one way to compensate for the terror of existence. Losing the *an da shealladh* over time, and through migration, might have been a sign that reality was becoming more bearable. One sight became enough to see by.

*

For several weeks, in the summer of 1967, a man worked in a hangar at the Royal Aircraft Establishment in Farnborough, Britain's chief aviation research facility, south-west of London. Patiently, but under considerable pressure, he was putting together the pieces of a broken passenger plane. The investigator, from the national Accidents Investigation Branch, was called Richard Clarke. He worked mostly alone. He was tall and slender, with an academic demeanour and a hawklike nose. He wore a suit and tie as he picked over the peeled wings, burned-out fuselage and densely cracked windscreens of the ruined cockpit.

The fragments of the aircraft, a propeller-driven British Midland C4 'Argonaut', whose call sign has been 'Hotel Golf', were laid on metal trestles and held in place by thin steel poles. Light came down through clear sections in the hangar roof above. Tables all around Clarke's skeletal reconstruction were covered with the fuzz of severed wires, seatbelts, light fittings, long shaft-like objects and miscellanies of bolts and bashed metallic scraps that he was yet to return to their proper places. Outside the hangar, the wreckage of dozens of earlier British air crashes, heaped any old how, lay open to the weather.

Hotel Golf had come down in the town of Stockport, a few miles from Manchester airport, shortly after ten o'clock on a Sunday morning in June. The plane had been carrying holidaymakers on their way back from Palma, in Mallorca. A few minutes before the aircraft had been due to land, it had mysteriously lost power. In the space of

145

twenty seconds, two of its four engines cut out. The pilot, Captain Harry Marlow, reported 'a little bit of trouble' and suggested that he make a turn around the airport. But the plane continued to slow and fall. It dropped below the clouds. People on the outskirts of Manchester looked up on a drizzly morning and saw the passengers waving down at them. Six miles short of the runway, Hotel Golf was only two hundred feet above the ground. The Argonaut crashed into a patch of industrial wasteland known as Hopes Carr, one of the few sparsely inhabited parts of Stockport. Seventy-two people died. Marlow was one of twelve survivors. He was pulled from the cockpit with head injuries and a broken jaw. Afterwards, the captain had no memory of the final minutes of the flight but in hospital kept asking: 'Which engine was it?'

The investigator looked for signs in the pieces around him. There was some urgency. The Argonaut was a Canadian variant of the Douglas DC4, which was in service around the world. Almost a thousand other airliners shared many of Hotel Golf's components, including its fuel system. What might go wrong in one could go wrong in any other. Clarke pulled at damaged wires and tested the plane's battered instruments. He found things that were 'non-plussing'. Some leads had been plugged in wrongly in one part of the plane and then plugged in wrongly at the other end as well, cancelling out the problem. There were anomalies and red herrings in the wreckage, informational noise that obscured the meaningful facts. Which was the part that signified? Where was the eloquent fragment that could tell the necessary story? Clarke noticed that the plane's rudder tab had been set at twelve degrees, indicating that the pilot had been straining to keep the aircraft level. He concluded that the fire that broke out after the crash had burned more fiercely on the starboard side, suggesting that those engines might have contained more fuel.

Clarke never solved the problem. In the autumn, two British Midland pilots, acting on a hunch, realised that it was possible for a pair of levers in the Argonaut's cockpit, which controlled the flow of fuel from one engine to another, to look closed when in fact they were partly open. Before the accident, the airline's crew believed it was impossible to inadvertently transfer fuel between engines during a flight. The public inquiry into the Stockport crash, which was held that winter, found that five days before the accident, Hotel Golf had arrived in Mallorca with just a few gallons left in one of its engines, after it had been mistakenly drained. The pilot had rejected the fuel measurement as incorrect and never mentioned it to anyone. The investigation concluded that on the plane's return to Manchester, its engine number four had run out of fuel. Then engine three stopped working as well – possibly because Captain Marlow shut down the wrong one ('Which engine was it?') and when he realised this, he didn't have time to turn it on again.

Witnesses on the ground described Marlow cutting the plane's power entirely in order to hit the sloping 'pocket-handkerchief' of wasteland where the plane came down. Hotel Golf narrowly cleared Stockport's hospital before crashing within a few hundred yards of its police station, town hall and an apartment block. The plane hit an electricity substation and a three-storey garage full of cars. No one on the ground was hurt. There was a large tear in the fuselage and the first rescuers on the scene, who included a Salvation Army band, came across a confused heap of dead and injured passengers, some torn from their seats, others still strapped in. There were many young families. Toys lay around. Fires broke out. It took an hour and a half to clear the dead and injured. During the day, an estimated ten thousand people converged on the crash site to gawp and pull at the wreckage. Photographs of the disaster showed the Argonaut's tail fin, emblazoned BM for British Midland, balanced precariously against a set of railings.

On 1 May, thirty-four days before the crash, Hencher had telephoned Barker, warning of another air disaster. The seer had wondered if he was suffering after-effects of his Cyprus premonition, but this vision appeared to him quite different. 'The plane has sweeping tail fins,' he said. Hencher said that the crash would happen within three weeks but he did not know where. More than sixty people would die. 'There are going to be some miraculous escapes and a number of survivors – I don't know how many. I can feel a lot of emotion concerning this. I feel a great sadness,' he said.

Barker asked Hencher about the potential victims. 'From the feeling I get there may be a lot of children involved,' Hencher replied.

Barker logged the call at 9 p.m. He passed on Hencher's message in a letter to Fairley the following night. It was just ten days after the crash in Cyprus and Hencher's warning about his own safety. It was late and the psychiatrist's letter was breathless. He seemed dazed by the importance, and the impossibility, of the information he had been given. 'Just as I am about to turn into bed it is, indeed, a terrible thing to reflect that there are now only two people in the world who know that some sixty people will die in a plane crash in three weeks' time,' Barker wrote. 'What are these people doing now? If only I could warn them somehow. Are they British?'

He promised to call Hencher back and press him about the distinctive tail fin, in case there was something that he had missed. The following month, the photographs from Stockport would convince Barker that Hencher had foreseen the Argonaut crash as well: the miraculous escapes, the children, the tail fin, the sadness.

'What if Mr. Hencher is right again?' Barker wrote to Fairley that night. 'But how can we stop it? If we could then Mr. Hencher would not be warned of this possible terrible tragedy in the way he was.'

There is no vision without a disaster to see. He was back at Jonah's quandary. He was as frustrated as he was excited. 'If only we could get more information,' Barker wrote. 'Perhaps someone else in England is like Mr. Hencher and has also been warned of this, but does not know of our scheme. If only we had more details. If only . . .'

FIG.

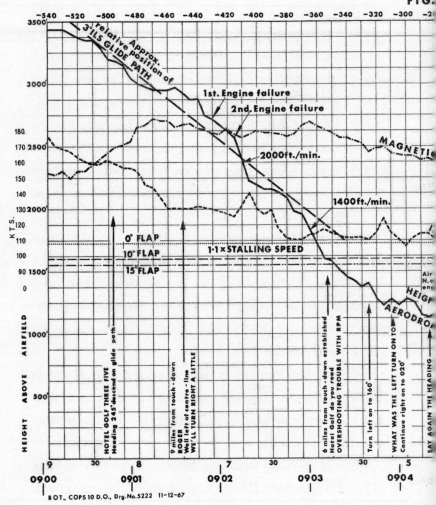

COMBINI

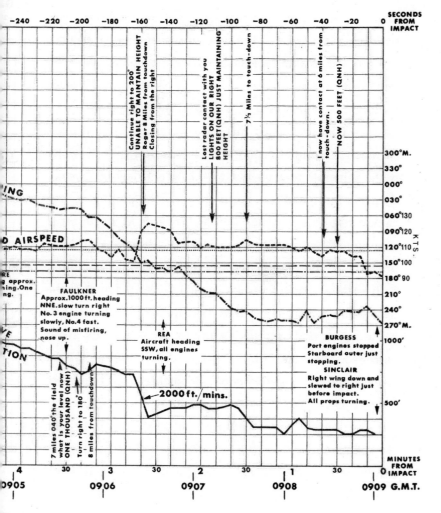

SECONDS FROM IMPACT

-240 -220 -200 -180 -160 -140 -120 -100 -80 -60 -40 -20 0

Continue right to 200°
UNABLE TO MAINTAIN HEIGHT
Roger 8 Miles from touchdown
Closing from the right

Lost radar contact with you
LIGHTS ON OUR RIGHT
800 FEET(QNH) JUST MAINTAINING
HEIGHT

7½ Miles to touch-down

I now have contact at 6 miles from
touch-down.

NOW 500 FEET (QNH)

300°M.
330°
000°
030°
060° 130
090° 120
120° 110 K T S
150° 100
180° 90
210°
240°
270°M.

ING

D AIRSPEED

RE
g approx.
ing.One
ng.

FAULKNER
Approx.1000 ft. heading
NNE. slow turn right
No. 3 engine turning
slowly, No.4 fast.
Sound of misfiring,
nose up.

REA
Aircraft heading
SSW, all engines
turning.

BURGESS
Port engines stopped
Starboard outer just
stopping.

1000'

VE

TION

7 miles 040°the field
what is your level now
ONE THOUSAND (QNH)

Turn right to 180°
8 miles from touchdown

2000 ft./mins.

SINCLAIR
Right wing down and
slewed to right just
before impact.
All props turning.

500'

MINUTES FROM IMPACT

4   30   3   30   2   30   1   30   0

0905   0906   0907   0908   0909 G.M.T.

ON PLOT

Time is decay. It can only go one way. In 1928 a British astro-physicist, Arthur Eddington, invented the phrase 'time's arrow' to describe its non-reversible nature. Eddington was building on the work of Rudolf Clausius and Sadi Carnot, a German scientist and a French engineer, who had observed that a proportion of energy is always wasted – dissipates to nothing – when heat is transferred from one object to another. Clausius called this measure entropy. He wanted the word to sound like energy, which was now gone. Entropy is the principle that underpins the second law of thermo-dynamics, which in turn is our only physical proof that time cannot run backwards as well as forwards. A single proof can sound flimsy until you consider the entropy taking place around you. It is the cup of coffee which cooled as you drank it. It is the energy of the sun which lit yesterday. It is the leaves that fall from the tree. It is the empire that fell. It is the emails that you don't reply to. Physicists talk about entropy as disorder within a system. A low-entropy system is well organised, tightly bound, all potential. It is life before time, Pangaea before the continents unfurled, a pack of eggs unopened in the fridge. Entropy is the release: the omelette, the cliffs eroding, the swallows migrating, the virus spread-ing, the asylum stagnating, chaos, climate change and death. 'The entropy of the universe', Clausius ruled, 'tends to a maximum.'

Entropy is human. We get it. Eddington was struck by how naturally the concept of time's arrow is grasped by our consciousness. Time flows, most of us agree, although

152

we do not know how. A story unfolds. My skin toughens and a worry line appears next to your eye. We do not grow younger. Some physicists have argued that the conceptual force of entropy, which underlies our own mortality, has made us believe that it encompasses more than it really does. In his 2018 book *The Order of Time*, Carlo Rovelli, an Italian theoretical physicist, posited the idea that the increasing entropy of the universe is, to simplify things greatly, just our way of looking at it. We may live in a small corner of existence, a tiny system, where the rule seems to hold. But what about elsewhere and other times? How do we really know that the past had lower entropy than the present, or that the future will not? We may only be seeing a fragment of the picture.

Part of our attachment to entropy is evident in our constant fight against it. We tidy the house. We elect new presidents. We dead-head the roses. We go to couples therapy. We try and put things back as they were, even when they will not go. Karl Friston, the neuro-imaging pioneer whose work has been central to the theory of the predictive brain, is deeply occupied by entropy. He sees reducing entropy, resisting disorder, as the goal of all forms of life. Friston calls this idea 'the free energy principle', a phrase also used by Freud, which expresses how our brains are always trying to conserve energy, and therefore unnecessary decay, in every mental process. Logic says that the more accurately we can predict the world, the less energy we will expend in coping and adapting to its vicissitudes, and the longer and more fruitfully we will live. 'Minimizing entropy corresponds to suppressing surprise over time,' Friston wrote in 2009. The free energy principle drives our memories, intuition and expectations to generate the smoothest experience of reality as it hits us. More than that, it governs our actions as well, as we struggle to bend that reality, to manage it and make it conform to our cognition. The entropy of the universe will not decrease. All of us must die. But seeing things before they happen is how we, as mortal souls, can seek to slow down time.

*

Bryn Farm, outside the Shropshire village of Nantmawr, lay about seventeen miles west of Shelton Hospital, almost at the Welsh border. On the morning of 25 October 1967, Norman Ellis, the farmer, called the vet about his pigs. Some of his herd were lame. Ellis had noticed the problem a few days earlier and wondered if the animals had arthritis. But when a veterinary officer arrived that morning from the Ministry of Agriculture, he diagnosed foot and mouth disease. Foot and mouth is one of the most contagious and economically damaging farmyard plagues. Sheep, cows, pigs, goats and deer – anything with a cloven hoof – can catch the virus. Animals normally survive but they are often weakened for life. The disease was first identified in the 1540s by a scholar from Verona, named Girolamo Fracastoro, who also gave the first full medical description of syphilis, in a poem that was 1,300 verses long.

Ellis kept sixty-seven pigs at Bryn Farm and all of them appeared to be infected. The vet signed a Form C, which stopped all animal movements within a five-mile radius. It was a Wednesday, which was market day in the town of Oswestry, four and a half miles away. Oswestry was in Dr Enoch's patch. Two of Ellis's cows were already there, along with 3,297 other animals for sale that day. The market was halted and the livestock impounded. One of Ellis's cows had already been sold, but a police car managed to flag down the lorry and bring the animal back to Bryn Farm. Throughout the day, the other livestock in Oswestry were inspected, found free of infection and let go.

There was no immediate reason to panic. The cases at Bryn Farm constituted Britain's third foot and mouth outbreak of the year. An infection in Warwickshire the previous month had been contained to four properties. The day after the pigs were diagnosed, an auctioneer came, along with two vets, to value the rest of Ellis's livestock – eighty-nine cattle and 244 sheep – which were then shot and buried under the orchards. Ellis's wife served the visitors roast lamb. (Six weeks later, officials concluded that the virus had been brought to Britain in a shipment of 770 Argentinian sheep carcasses, one of which Ellis had bought.)

On the Saturday, the neighbouring farmer, who shared a pig weighing machine with Ellis, reported foot and mouth. By Monday 30 October there were nine cases – one of which was a hundred miles away in Lancashire. Two days later, there were nineteen. The numbers spiralled after that. In late November, when the epizootic peaked, eighty-one farms reported fresh foot and mouth infections on the same day, across two thousand square miles of English farmland. The Cheshire plain, which stretches from the Welsh border almost as far as Liverpool, had the highest density of dairy cattle in Britain in the sixties. It became a landscape of pyres. The army was called in to help shoot half a million cows, sheep and goats. There was a shortage of steel chains to drag the corpses around, so RAF helicopters shared them from farm to farm, hovering over fields and dropping them from nets. Smoke from the burnings drifted in the autumn winds.

The virus reached the farm next to Barker's house. His children saw the farmer laying down disinfected straw at the entrance and heard gunshots in the fields. The outbreak that began with Ellis's pigs was Britain's worst foot and mouth infection of the twentieth century. The countryside was stilled. Shrewsbury and Shelton were at the heart of it. Hundreds of the hospital's patients and staff came from rural communities where ordinary life was upended. Schools were closed. Footpaths were sealed.

Inside the hospital, Barker struggled against a feeling of doom. The previous month, he had completed an eighteen-month study of his older chronic patients at Shelton, to see if he could improve their lives. On Thursday mornings, he and Mabel Miller, his aversion therapy researcher, had toured the crowded back wards, intervening in the treatment of 116 men and women: modifying their medication, if they were taking any; attempting to contact their fam-

ilies, to arrange visits; applying for transfers to halfway houses or hostels, if they were ready to return to normal life. After a year and a half of intense effort, thirty-four of his patients were no longer in the hospital. Eighteen had died. Of the sixteen that the psychiatrist had managed to discharge, only two had gone home. Most of the patients' families wanted nothing to do with them. Over the same period, Barker had been forced to take on another eighteen chronic patients.

Barker's findings confirmed a feeling of desperation, and anger, among doctors and reformers who wanted to transform Britain's mental hospitals. Six years earlier, in the spring of 1961, Enoch Powell, then the minister of health, had announced that it was time to demolish the country's

Victorian asylums. 'This is a colossal undertaking,' Powell told a conference in Brighton. 'There they stand, isolated, majestic, imperious, brooded over by the gigantic water-tower and chimney combined, rising unmistakable and daunting out of the countryside – the asylums which our forefathers built with such immense solidity to express the notions of their day.'

Powell wanted the places physically as well as conceptually destroyed. 'We have to get the idea into our heads that a hospital is a shell,' he said, 'a framework, however complex, to contain certain processes, and when the processes change or are superseded, then the shell must most probably be scrapped and the framework dismantled.'

The moral and intellectual case for reform was unarguable. Since 1955, new antipsychotic drugs had helped to transform the treatment of the most disturbed. For the first time, it was possible to imagine that the vast majority of the mentally ill could be treated either as outpatients, or in general hospitals or small clinics close to their homes. Doctors had also demonstrated the harm caused by long-term incarceration itself. In 1959, Russell Barton, a psychiatrist at Severalls Hospital in Essex, who had helped to treat survivors of Bergen-Belsen concentration camp, used the phrase 'institutional neurosis' to describe the affliction suffered by many chronic patients. Barton identified seven causes of the condition, including enforced idleness and 'the loss of personal events'. He observed that sufferers gave up on the idea of the future.

But the plan to close the nation's asylums, or dramatically shrink them, had to overcome the reality of the institutions themselves. During the summer of 1967, Barbara Robb, a psychotherapist and mental health campaigner, published *Sans Everything*, a grim account compiled from anonymous patients, nurses and doctors that showed how little was changing. *Sans Everything* was inspired by Robb's attempts to help a former client, Amy Gibbs, who had been admitted to Friern Hospital, a two-thousand-bed asylum in north London, in 1963. Gibbs was an artist and seamstress in her early seventies who was referred to the hospital after suffering from anxiety and side effects from her medication. But once she was on a back ward, her mental health deteriorated rapidly. For much of 1965, Robb visited Gibbs twice a week, sometimes carrying a cassette recorder in her handbag, trying everything in her power to get her discharged.

The book documented a world of institutional neurosis, in which elderly patients lived sixty to a ward, deprived of their dentures, glasses, hearing aids and selves. The nursing staff that Robb saw weren't overtly abusive, but they were overstretched and short-tempered. They treated the patients as children. They gave themselves every excuse not to care. Robb formed a campaign group, AEGIS (Aid for the Elderly in Government Institutions), which she ran from a cottage in Hampstead. Robb was fifty-five, well connected and driven. She carried a hint of magic. She was from an old Catholic family and claimed to have 'the

159

blood of six martyrs in my veins'. In 1949, on the night before the Epsom Derby, Robb dreamed of the winning three horses, in the correct order. She trained as an analyst and had travelled to Vienna to meet Jung, who found her 'Quite remarkable. If ever there was an anima, it was she.' Robb's outspokenness and public profile made her a figure of hope for doctors and nurses, who were trapped in slowly changing hospitals and felt the same way that she did. Enoch was one of Robb's allies. He contributed a chapter to *Sans Everything*, based on his experiences at Shelton. When Barker completed his patient survey, he sent a copy to AEGIS.

Foot and mouth was found among the animals on Shelton's farm in early December. On 4 December, all of the hospital's livestock – sixty-eight cattle and 234 pigs – were killed. In a way, the slaughter marked the end of the asylum era at Shelton. The farm never recovered, and was sold. The hospital could no longer function as a world unto itself. Ten days later, Barker gave a lecture in Shelton's main hall about the long-stay population. Littlejohn and the rest of the consultants came, along with regional health officials. Barker asked for the lights to be turned out, and he showed slides of the old men and women whom he had encountered on his ward rounds. 'Here were two gentlemen who were rather non immaculate,' the psychiatrist said, showing old men with food stains down their fronts. Others were missing ties and pieces of clothing. Not all of the images were disheartening. Barker showed a woman in ward seven who was being carefully tended by nurses. 'Very dour looking, very pale, very fragile looking,' he said. 'But the sort of patient that we have to care for.'

He was by turns dispassionate and tender. According to the Ministry of Health, Shelton Hospital was due to have 448 beds in 1975. In the autumn of 1967, that meant a reduction of forty per cent of its patients in the space of eight years. But Barker pointed out that the hospital's senior nurses and administrators were still paid according to how many beds were occupied, meaning they had little incentive to empty the wards. 'We've got to, in fact, keep the numbers up from this point of view, rather than to reduce the numbers,' he told his colleagues, to silence. 'One just mentions that point in passing.'

Barker put up slide after slide of statistics: sixty-nine per cent of chronic patients had been at Shelton for more than five years; fourteen per cent of female patients did industrial therapy. Seventy-seven per cent of the chronic patients came from Shropshire's countryside, rather than its towns. He mused out loud, enjoying some of the findings for their own sake. 'Whether it means that rural patients in our area tend to become more easily mentally ill . . . one doesn't know,' he said. 'I thought it was interesting.' He found that unmarried patients received the fewest visitors. 'Which is what you'd expect,' Barker said. 'But it's nice to find out these things for oneself.' At one point, his torch went out. He ended his lecture admitting that he neglected older patients in his own work. 'But for many chronics, it is seen that our hospital will be their only hope. They will live out the remainder of their lives and die here,' Barker said. 'And we must therefore make it as pleasant as possible for them . . .' He searched for the right words. 'I mean, as if they were our relatives.'

Afterwards, there was a discussion. Enoch, who had staged the lecture as part of his education course, was shrill and pointed. 'We've got to answer the question: what is this hospital for?' he demanded. Littlejohn, as usual, barely said a word. One of the older doctors, a Dr Thomas, recalled working at a psychiatric hospital in Abergavenny and finding that the rates of death and discharge were roughly the same 120 years earlier. 'It was very humiliating,' he said. 'I don't think the Ministry is aware of it.'

Barker and Enoch, the young reformers, dominated the conversation. 'I hope you're going to pay me for what you use in your article,' Enoch said, ribbing Barker in front of Littlejohn for his taste for publicity, which had caused so much trouble. 'Oh, straight into the *News of the World*,' Barker replied. 'Straight to the *News of the World*.'

Talking with his colleagues, Barker acknowledged, again, his own malaise when faced with a ward full of bereft, motionless long-stay patients, who probably should not be in the hospital but for whom there was no other place. 'It's so easy to overlook these people. You might do a ward round and you say, "Right, well, I'll leave this ward, there won't be anything there." And one can get into the habit, and you can leave it for months and months and years and years. I know myself,' Barker said. 'This was done to help me visit them, perhaps.'

The conversation turned back to the patients' clothes and appearance, which Barker thought had a vital effect on everyone's morale.

'Which is going to die first?' Enoch chimed in. 'All the chronics we've got now or these old thick suits?'

'The chronics,' Barker said. He didn't miss a beat. 'The suits will go on.'

*

The Premonitions Bureau offered him a way out. Every day, Jennifer Preston received a vision or two, through a letter or a call to the newsroom. During the course of 1967, the bureau received 469 warnings, most of which proved impossible to verify. But there were some 'best buys', as Fairley liked to call them.

On 22 May, Lynn Singh, from Thetford in Norfolk, described a dream she had had the previous night of 'a tremendous conflagration at a considerable distance'. Singh had seen a large building with 'huge girders', flames leaping one hundred feet high and 'sudden brilliant flashes every few seconds, which would seem to indicate mighty explosions'. Around the same moment that Singh posted her letter in East Anglia – at 2.45 p.m. on a Monday – the first fire crews reached L'Innovation, a huge art deco department store on the Rue Neuve in Brussels, where flames had soared up a

163

high central atrium to a glass roof supported by arching iron beams. Two hundred and fifty-one people lost their lives in the fire, which consumed the building within minutes and remains Belgium's worst peacetime disaster. The fire flashed and raged, fed by butane gas tanks exploding in the camping department.

Two weeks later, on 5 June, the Egyptian army blocked both ends of the Suez Canal, marking the start of the Six-Day War with Israel and fulfilling a prophecy made by Michael Sadgrove, of King's Lynn, who had dreamed of 'a merchant ship stuck in Suez Canal about the end of May'. (Fifteen cargo vessels were trapped in the canal for the following eight years.)

Fairley remained tantalised by the betting possibilities. The 1967 Grand National was won by Foinavon, a 100–1 outsider, after an extraordinary pile-up at an innocuous fence. The day after the race, a young Australian named George Cranmer called the bureau and claimed that he had dreamed of Foinavon's colours the night before and knew that the horse was going to win. Fairley invited Cranmer to call again. On the morning of the Epsom Derby, two months later, a friend of Cranmer's called and put him on the line. Cranmer was nervous, but he said he had dreamed of a set of jockey's colours and had seen a vision of the winning horse being led away. This time the colours belonged to Ribocco, a 20–1 shot. 'I did the same before Foinavon won the National,' Cranmer said. Ribocco came second in the Derby, but won the Irish Derby the following month, in front of a watching Jackie Kennedy. 'Premonition?' Fairley wrote in the *Standard*. 'Or coincidence?'

Barker and Fairley continued to drum up interest in the project. In September, Barker went to London to take part in a telepathy experiment, organised by Alister Hardy, a professor of zoology who later founded a Religious Experience Research Unit at the University of Oxford. Hardy had become fascinated by ESP while serving in a cyclist battalion during the First World War. He was posted to Lincolnshire, on the eastern coast of England, where he befriended the widow of an early member of the SPR, who seemed able to read his mind. In the autumn of 1967, after he had retired from a distinguished career studying whales and plankton, Hardy staged an ambitious public experiment at Caxton Hall, in Westminster. Over the course of seven Mondays, Hardy stood on a stage and drew large chalk sketches or projected photographs, which a crowd of two hundred people then tried to mentally convey to twenty test subjects who sat in cubicles sealed with blackout material. The percipients then drew whatever images they received from the ether.

On 18 September, Barker sat in one of the blacked-out cubicles. He reckoned he received only two out of ten images, which he was disappointed by. 'My own performance was rather poor,' he wrote. But he was impressed by the professionalism of the research. (Hardy concluded that there were only thirty-five 'hits' out of 2,112 responses during the whole experiment.)

Barker was also in touch with Brian Inglis, a former editor

of *The Spectator*, who was becoming one of Britain's leading proponents of the paranormal. The men exchanged letters and Barker suggested that they meet for dinner on 6 November. He was due in London that day to address a meeting at the Royal College of Medicine about his work on Aberfan. He planned to travel from Shrewsbury that morning.

A few days earlier, Miss Middleton sent an unusually explicit warning to the bureau. On 1 November, the music teacher had found herself feeling acutely depressed. She sat in her kitchen in Edmonton. 'Gradually I saw a streak, then a flash of light and then a sort of grey mist. I was trying to find out where it was,' she said later. 'The word "train" kept coming through. Train . . . train.' Miss Middleton put her vision in a note to the bureau: 'I see a crash . . . maybe on a railway . . . a station may be involved . . . people waiting in the station and the words Charing Cross. The sound of A CRASH.' On 11 October Hencher had also written, warning of a mainline rail crash, in which many people would be killed and two carriages would come to rest one on top of the other.

The night before Barker's meeting in London, on 5 November, the 19.43 express service left Hastings, on England's south coast, heading for Charing Cross. It had been a mild day on the seaside but the evening had turned cold and wet. The train was busy with returning day trippers and office workers who were heading up to the capital for the week. By the time the service reached Sevenoaks, twenty-two miles from Charing Cross, and switched drivers, it was running four minutes late

168

and there were passengers standing in the corridors. The new driver was called Donald Purves. He had driven the same route two hours earlier, on the 17.43 service, and had worked the line for nine years. As the train clattered through Grove Park, a small suburban station in southeast London, Purves prepared to apply the air brake to slow from 70 to 60 mph, to comply with the city's speed restrictions.

When he reached for the brake, Purves felt a momentary drag on the 750-foot train, which weighed almost five hundred tons. He wondered if he had engaged it too soon, and gave the brake a moment to sort itself out. The drag that he felt was the train derailing. A set of wheels at the front of the third carriage had hit a five-inch break in the tracks and jumped the rails. Albert Green, a signalman at Hither Green, the next station, saw a shower of sparks coming from the bottom of the train. Passengers on board thought it sounded like they were running over glass. Others described the sound of rocks hitting the carriages. A guard named Gray stuck his head out of a window in the sixth carriage and had his hat knocked off. He shouted at people to get on the floor. The train stayed upright for 463 yards before the derailed wheels hit a junction and flipped four carriages on to their sides. People were thrown about like dolls. Two of the carriages had their sides ripped off. Windows smashed all the way down the train. Roofs were crushed. Chairs, luggage, newspapers and coffee cups were thrown on to the track. 'Everything happened so fast,' Purves said. In the engine car, the driver felt the whole train rear up and then heard an enormous bang. The emergency brake kicked in. His coach came to a halt 220 yards down the track, on its own. When he climbed down, the rest of the train was gone.

The first ambulances arrived six minutes later. Police officers, fire crews and local residents scrambled up the embankment. The wreckage lay across a bridge over St Mildred's Road, just short of Hither Green station and about eight miles from Charing Cross. The rescuers brought ropes to guide survivors down the dark bank, which was slippery in the rain. Blue lights from the emergency vehicles and sparks from the firefighters' angle grinders lit the rails. Forty-nine people had died, although at first the death toll was thought to be much higher. At the time of the crash, Alan Hencher was on a shift at the GPO. He complained of a severe headache and was taken to the sick bay. At 10.15 p.m., he wrote a note saying that he thought there had been a railway accident and that it might have happened about an hour ago. The train had come off the rails at Hither Green at 9.16 p.m.

Miss Middleton's 'Charing Cross' prediction and Hencher's turn in the sick bay put the Premonitions Bureau back in the headlines. When he came up to London to deliver his lecture the next day, Barker gave an interview to the BBC. The *Evening News*, the *Standard*'s great rival, ran the 'Strange Case of the Two Who Knew' on its front page, alongside coverage of the crash, which was Britain's worst railway disaster for a decade. 'I have talked to these two shaken people today,' Michael Jeffries, the news-paper's science correspondent, wrote of Hencher and Miss

Middleton. 'Somehow, while dreaming or awake, they can gate-crash the time barrier . . . see the unleashed wheels of disaster turning before the rest of us.' Barker told Jeffries: 'They are absolutely genuine. Quite honestly, it staggers me.'

*

Seventy-eight people were injured in the crash. Among those slightly hurt were a teenage couple who had been sitting in first class. The boy had long hair, prominent teeth and elfin eyes and was wearing a mackintosh and a trilby; the girl was wearing a green-and-white coat with a fur collar.

The boy was Robin Gibb, the seventeen-year-old singer for the Bee Gees, with his fiancée, Molly Hullis, who was a receptionist for the band's manager, Robert Stigwood. Gibb had flown into London that morning from Berlin and had spent the day with Hullis and her parents in Hastings, who had given them a bread pudding and some apples for the journey. Hullis joked that Gibb was travelling so much he should take out travel insurance. When the train began rocking, she reassured him that it was always like this coming into London. The express trains went too fast. But Gibb didn't believe her. He stood up to pull the emergency cord, which was when the lights went out and their carriage overturned. 'Big stretches of railway line came crashing in straight past my face,' Gibb told a reporter the next day. 'It almost seemed as if the train was going to pieces around us. One minute we were in the luggage rack, the next we were

171

on the floor.' Gibb helped Hullis through a broken window. His hair was full of broken glass. There was oil on their clothes. They walked along the top of the carriage, pulling survivors through the broken windows. There were terrible screams. 'All this to get to Battersea Funfair,' Gibb joked. It was Guy Fawkes night. Fireworks blazed in the rain.

Afterwards, Gibb put his and Hullis's survival down to having the money to travel first class. Their carriage had a corridor, full of standing passengers, which took the main force of the impact. The Bee Gees had had their first hit in the spring of 1967, with 'New York Mining Disaster 1941'. Robin wrote the song with his elder brother, Barry, in a darkened stairwell of Polydor Studios, on Stratford Place, near Bond Street, one night in early March. It was about a miner trapped underground, waiting to be rescued, and was inspired by the Aberfan disaster four and a half months earlier. 'That song didn't take a lot of thinking about because it is a catastrophe and catastrophes happen all the time,' Barry said later. 'The atmosphere just came and the song just came.'

The Gibb brothers were born on the Isle of Man but grew up in Australia. They had sailed to London in January that year to become pop stars. 'New York Mining Disaster 1941' was their first single and reached number fourteen in the US charts, partly because many radio stations and listeners believed that the Bee Gees were really the Beatles in disguise. Promotional copies of the single were distributed with a blank label and the information that it was by a new British band beginning with B. (There was a rumour that 'BG' stood for 'Beatles Group'.) The song was odd and somewhat haunting. 'In the event of something happening to me,' Robin sang. The title referred to nothing. There wasn't a New York mining disaster in 1941. DJs played it with vague, cryptic introductions. No one was quite sure what they were listening to. People heard what they wanted to hear.

IV

One day in 1995, in the German cathedral city of Mainz, a fifty-one-year-old woman went to hospital to undergo a procedure on a tumour that was growing in the base of her skull. Frau K was a polite, small woman who seemed prematurely aged by a complicated medical history. She had first undergone surgery on her tumour sixteen years earlier and had been hospitalised many times over the years. As a patient, she was reserved and made little impression at a meeting with her physician, a forty-four-year-old neuroradiologist named Wibke Müller-Forell, a couple of days before the operation.

'She was a nice, lovely . . . in my opinion, old little woman,' Müller-Forell recalled. The procedure was an embolisation of the blood vessels supplying the tumour. During an embolisation, a very narrow tube is fed into an artery in the patient's leg or back, and then threaded carefully up, through the narrowing channels of the circulatory system, until it is in the right place, where tiny plastic particles are then used to block up the blood vessels. Stopping the flow of blood to a tumour makes it easier to remove at a later date. Embolisations are routine and are normally carried out under sedation rather than general anaesthetic. The word embolism has come to mean a blockage, normally in the body somewhere, but the term originally described the adding of days to the calendar to align the twelve lunar months of the year with the Earth's 365-day orbit of the sun. It is an intercalation. A correction of time.

The truth was that Frau K was uneasy. She did not want to have the embolisation. She was tired of hospitals and of surgery. She had lived with the tumour for a long time and she wasn't in pain. It was her husband who persuaded her that the intervention was necessary, even though Frau K confided to him that this time she feared she would not survive. Müller-Forell and the rest of the team at the Neuroradiological Institute of Mainz's medical school did not know any of this. On the day of the procedure, Frau K just seemed unusually anxious. She kept asking how long the embolisation would take and if it was going well so far. 'Very, very, very busy,' Müller-Forell recalled, of the patient. It was hard to soothe her, either with words or with medication. The anaesthesiologist noted an 'emotional imbalance' in Frau K and gave her midazolam, a powerful sedative that would, among other effects, temporarily stop her brain from creating new memories. She was also given methimazole to reduce the likelihood of a 'sympathetic storm' – the term used by Walter Cannon, the investigator of voodoo death, to describe how stress hormones, known as catecholamines, can overwhelm the heart.

Frau K was drowsy, almost vegetative, when the embolisation began but her dread lingered in the operating room. During her career, before Frau K and after, Müller-Forell had certain patients who were convinced, whatever she had to say about the actual danger of their illness or their sincere regard for her skills as a physician, that they had entered the final phase of their lives. Most of them were perfectly civil about it. Sometimes they were right. Frau K was one of these. 'Her existential fear of dying at any time during the planned procedure predominated,' Müller-Forell noted in a later case report.

When Müller-Forell inserted a catheter into the left vertebral artery, Frau K, although sedated, gave a deep, sudden groan and lost consciousness. An aneurysm had ruptured in her brain. Survivors of subarachnoid haemorrhage describe the pain as percussive. English medical phraseology speaks of a thunderclap inside your head. In German, it is a *Vernichtungskopfschmerzen*, an annihilation headache. An emergency CT scan and angiography revealed extensive bleeding in Frau K's brain and into her spinal column. She died two days later. At first, Müller-Forell and the rest of the medical team assumed they had done something wrong. Doctors are used to being the protagonists, for good or ill, in dealings with their patients. But the autopsy showed that there had been no errors during the procedure. Frau K was not a well woman but, based on the medical evidence, it was her terror that killed her. When Müller-Forell published her case report in 1999, she called it: 'Psychic Stress as a Trigger of the Spontaneous Development and Rupture of an Aneurysm?'

The influence of negative expectation – of fear – on our health is known as the nocebo effect. The term was first used by Walter Kennedy, a British doctor and public health expert, in the early sixties, to describe the opposite of the more benign placebo. (*Placebo* means 'I will please,' in Latin; *nocebo* means 'I will harm.') Kennedy served as a colonel in medical intelligence during the Second World War. He translated German scientific papers and attempted to glean the secrets of Nazi military medicine. After the war, he became the principal medical officer of Distillers Company, a British drinks business that had branched into pharmaceuticals. In the summer of 1956, Kennedy returned from a research trip to Aachen, where he had visited Chemie Grünenthal, a German drug manufacturer, which had recently developed an exciting new sedative called thalidomide. Kennedy tried the drug himself and found it excellent for his asthma. He recommended that Distillers license the drug for the British market as soon as possible.

Kennedy used the term 'nocebo effect' to describe the vague, non-specific negative reactions that occurred among patients, particularly during drug trials, which did not appear to have a rational cause. 'It refers to a quality inherent in the patient rather than in the remedy,' he wrote. Kennedy never claimed it was a particularly original idea. 'Every doctor has met the nocebo reactor,' he wrote, 'even if he has not labelled him as such.' Kennedy thought it might be difficult to study nocebo effects on any significant scale because they would be specific to individuals. He compared them to the content of dreams. He viewed the phenomenon as an interesting minor irritant – a distraction from the real work of making and introducing new medicines. Kennedy wondered how many beneficial drugs had been discarded during trials because of dubious or psychosomatic reactions among the test subjects.

In Britain, thalidomide was licensed to cure morning sickness with 'absolute safety' even though the drug was never tested on pregnant women. Neither the placebo nor the nocebo effect came into it. The drug went on sale in the spring of 1958. Kennedy's paper on the nocebo effect appeared in September 1961. Two months later, a Distillers representative in Sydney was warned by an Australian obstetrician named William McBride that mothers in his clinic who had taken thalidomide were giving birth to babies with webbed fingers and extra toes, missing hands and feet, or dramatically shortened legs and arms, a condition known as phocomelia. Thalidomide was withdrawn from sale in Britain two weeks later.

In its three years on the market, thalidomide affected thousands of pregnancies and babies. Untold numbers of foetuses were lost. Kennedy retired to Scotland soon afterwards, where he studied teratology, the development of malformations, deep into his old age, in the hope of finding an alternative explanation for the disaster.

The nocebo effect is still observed most commonly during drug trials, where self-fulfilling prophecies come to pass. Volunteers warned about possible side effects frequently develop them, even if they have been taking inert sugar pills. In 2005, 120 Italian men who were being treated for enlargement of the prostate gland were given a drug called finasteride. One group was told that the drug could cause sexual problems, although this was uncommon, while the other group was not. After a year, forty-four per cent of those in the first group complained of erectile problems and decreased libido, compared with fifteen per cent in the second. Each year, millions of people around the world stop taking statins, which reduce cholesterol levels, because of side effects that include fatigue, muscle aches, joint pain and nausea. In 2020, a study of patients who had given up statins found that they also complained of ninety per cent of their side effects when given a placebo. For a fact of life, the nocebo effect is hard to study. Most researchers don't like to induce suffering, even if they are allowed to. The response occurs for the most part at the edges of things, an anomaly recognised only after the fact, once more pleasing theories have been tried and failed.

It takes a certain kind of doctor to test the nocebo directly. In the summer of 1885, a thirty-two-year-old woman, 'very stout, well-nourished but physically weak,' was referred to Dr John Noland MacKenzie, a surgeon in Baltimore, with crippling hay fever and asthma. The woman was confined to her bed for weeks every summer and autumn. If a hay-cart passed her in the street, she would suffer a paroxysm. She could not touch a peach. Ordinary medicine provided no relief. Cold weather and trips to the seaside were somewhat beneficial. Cocaine relieved her symptoms for half an hour at the most, 'leaving her, as a rule, worse off than she was before its application.'

MacKenzie prepared a prescription of his own devising. After two weeks, he wrote, the woman felt much better. After a month, MacKenzie invited the patient to his consulting room, where he had hidden an artificial rose – 'a perfect counterfeit of the original' – behind a screen. MacKenzie had wiped every petal before she arrived, to make sure it was clean. He checked that the woman was well and then sat down in front of her with the rose in his hand. After a minute, she began to sneeze. Within five minutes, her nose was clogged and inflamed. 'The feeling of oppression in her chest began, with slight embarrassment of respiration.' Breathing became a struggle. 'As I considered the result of the experiment sufficiently satisfactory, I removed the rose and placed it in a distant part of the room,' MacKenzie wrote. When he told the woman that the flower wasn't real, she inspected it leaf by leaf, in disbelief, her nose streaming.

In 1968, a team of psychiatrists in Brooklyn asked forty asthma sufferers to help them with a study of air pollutants. Almost half experienced a tightening of their airways when they inhaled a harmless vapour of saline solution. Twelve had full-blown asthma attacks. The nocebo effect can stop good medicine from working. When researchers gave migraine sufferers rizatriptan, a powerful migraine medication, but said it was a placebo, it turned out to be half as effective. For about half an hour, patients with Parkinson's disease who have electrodes implanted in their brains to help smooth their movements will do better or worse at motor tasks depending on whether they believe their electrodes are working, rather than whether they actually are. During a placebo trial described in 2006, a twenty-six-year-old man swallowed twenty-nine inert capsules, thinking they were antidepressants, in an apparent suicide attempt. His blood pressure collapsed and he was taken to the hospital, where the symptoms abated when he was told what he had taken.

Researchers who study the nocebo effect draw a distinction between our conscious expectations and other more tenuous forms of anticipation which have less to do with our understanding. The nocebo effect is at least partly contagious. If you see someone suffering pain or entering an altered state during an experiment, you are much more likely to experience the same thing when it is your turn. The social force of the nocebo effect means that it can give rise to peculiar, localised conditions. In 2007, the maker of Eltroxin, a thyroid-replacement drug distributed in New Zealand, moved its manufacture from Canada to Germany. The active ingredients in the drug remained the same, but the new pills looked different. They were larger; some were off-white, rather than yellow. After the media reported that the new Eltroxin was cheaper to make, reports of side effects among New Zealanders rose by a factor of two thousand. Beliefs do not become weaker because they are shared. Between 1977 and 1982, more than fifty Hmong refugees in the US, who came primarily from Laos, died from sudden nocturnal death syndrome. Their hearts stopped in the night. The community generally interpreted these deaths as the consequence of lethal nightmares, known as *dab tsog*. People became afraid to go to bed. 'You can't help but behave in a way that is the result of having been immersed since birth in a certain set of attitudes and thoughts,' recalled Shelley Adler, the director of the Osher Center at the University of California, San Francisco, who interviewed hundreds of Hmong people about the deaths. Post-mortems revealed that some of the victims suffered from abnormal heart rhythms, which could have been exacerbated by the stress of immigration and the fear of an evil spirit crushing their chest in the night.

In the early 2000s, the children of families awaiting asylum in Sweden began to fall into mysterious stupors. Their parents, who were often from former Soviet republics, brought them to hospital emergency rooms when they had stopped eating, drinking and talking. Between 2003 and 2005, 424 cases of the new condition were reported. Caught up in bureaucratic sagas that they did not understand, moved endlessly between apartments and schools, destabilised, and seemingly blaming themselves for their families' predicaments, the children shut down. They could be kept alive by tube feeding and careful nursing, but doctors and psychiatrists were dumbfounded by their decline and the complex, forbidding knot of traumas that held them.

Resignation syndrome – *uppgivenhetssyndrom*, as it became known – caused great consternation in Sweden. Sceptics, who were often critical of the country's relative benevolence towards refugees, suggested that the condition was a form of malingering or even child abuse, designed to elicit a residency permit for the family. Doctors, child psychologists and health workers, meanwhile, pointed to the existential stress caused by growing up in a permanent state of helplessness and hopelessness (most of the families had been refused asylum several times) and the intergenerational psychological damage that these children carried.

The result was a nocebo effect. The children expected nothing and became nothing. People compared their catatonic forms to creatures from a fairy tale. They were often pre-pubescent, budding with life, yet frozen in time. Göran Bodegård, a child psychiatrist in Stockholm, described five cases of the condition in 2005. He was struck by the stifling desperation of the patients' mothers, who were convinced that their children were dying. 'This state of "Madonna dolorosa" brought about an atmosphere of "pietà", filling the hospital room with a heavy, claustrophobic feeling, which affected everyone who entered,' Bodegård wrote. 'One moved carefully, whispering, not talking directly to the child, whose presence one only could barely detect in the bed, lying immovable and covered by a sheet or quilt.'

In 2014, the Swedish National Board of Health and Welfare officially recognised resignation syndrome as a diagnosis and recommended that 'a permanent residency permit is considered by far the most effective "treatment"'. The children normally recovered slowly, within six months or a year of asylum being granted. More children fell ill. Between 2014 and 2019, another 414 cases were reported.

Karl Sallin, a paediatrician at Karolinska University Hospital in Stockholm, was raised in a conservative family that was doubtful about the reality of mental illness at all. About ten years ago, he began to wonder why *uppgivenhetssyndrom* was only being reported, and treated, in Sweden, when its sufferers came from all over the world and their experiences, while appalling, were not unique to Scandinavia. He found each of the main, vying explanations – malingering refugees or a brand new, complex illness – equally unsatisfying in its way. Sallin interviewed fellow doctors and psychiatrists and sat with families in silent rooms, next to their motionless children. Sometimes he felt in contact with something much bigger than himself.

In 2016, Sallin wrote a paper in which he described resignation syndrome as 'culture-bound', a phrase that originated in the sixties to describe forms of psychosis whose causes were primarily social, animated by shared beliefs, rather than anatomical or stemming from an individual's state of mind. Instead of being a concocted illness or a new discovery by the Swedish medical establishment, Sallin suggested that resignation syndrome was a combined creation of everyone involved, what psychologists sometimes call an idiom of distress.

In the nineteenth and first half of the twentieth century, doctors used to recognise and treat 'compensation syndrome', symptoms that normally developed during lawsuits or claims for damages of some kind (it was also known as 'railway spine' and 'profit neurosis') and which did not always dissipate when a verdict was reached. Sallin also theorised that resignation syndrome in Sweden could be understood through the model of the predictive brain: in this interpretation, the children had such low and fragmented expectations of the future, such strong priors, informed by their lives and the shattered existence of their parents and former societies, that their minds and bodies entered a kind of feedback loop of despair, which was then unwittingly confirmed by sympathetic health workers, who believed that the only way they could be cured was if the state granted asylum to their families.

'We actually induce different ways of being ill by treating it in certain ways,' Sallin said. 'If you set up the residency permit as a treatment, that will actually create an illness, and so on and so on.' He found the shared fixations that could be constructed between powerful figures, such as psychiatrists, and vulnerable ones, such as catatonic, wraith-like children and their crazed parents, to be both worrisome and inescapable.

Sometimes Sallin found himself thinking about Henrik Ibsen's play *The Wild Duck*, in which a misguided idealist, Gregers, sets about stripping a fragile family of its illusions so that each person can see the truth of their situation. Gregers's mission ends unbearably. Ibsen calls the fictional version of reality that each one of us knits, moment by moment, within ourselves our *livslögnen*, or life-lie. Do we

186

even exist without the meanings that we invent for ourselves? 'We all carry lies that we can't get rid of,' Sallin said. 'I shiver when I think of that play.'

*

On 7 February 1968, two months after the Hither Green rail crash, Miss Middleton saw a vision of Barker. On one side of the apparition, there was the doctor's head and shoulders: his hair was thinning now; his sideburns were somewhat wild and grey; his eyes were pale brown. On the other side, Miss Middleton's parents, Henry and Annie, stood waiting. A path ran between the figures. There was a glimpse of a boy and a girl. The image persisted for a week. Miss Middleton's parents had been dead for five years. They often occurred in her dreams, or otherwise appeared to her, and she associated their presence with misfortune. 'Not wishing to alarm anyone . . . I merely said that my parents were trying to tell me something,' Miss Middleton wrote later, in her memoir. 'I interpreted this as something concerning the doctor.'

Scared to Death was published later that month. In the book, Barker described the cases of forty-two people – twenty-eight men and fourteen women – who appeared to have succumbed to fear in a variety of situations, from exotic curses to plain shock or a slowly building sense of hopelessness. He took his examples from concentration camps during the Second World War, the Labrador case from 1965, reports from British colonial doctors working in sub-Saharan

Africa, and patient histories closer to home: the headmaster of an art school, who died suddenly after a successful glaucoma operation because he believed he had lost his sight; a seventy-eight-year-old businessman whose usual doctor had decided not to tell him that a small lump on his tongue was cancerous, and who died a few days after it was diagnosed by somebody else. Barker wrote for a mass audience, presenting himself as an uncompromising investigator. 'I did not see that we could just leave it at that,' he wrote of the Labrador case. He explained the work of Cannon and Richter's rat experiments, and advanced his own analysis of why some presentiments of death come true and others don't. 'It is necessary to consider the seed and the soil,' Barker concluded. The psychiatrist ascribed the potency of a prediction to its 'sense of inevitability', the personality of the person who receives it, and the way it interacts with our deepest beliefs about illness and death.

The book also made riskier disclosures. Over the warnings of Littlejohn and the regional NHS officials, as well as a second lawyer, whom Barker contacted a few months before publication, he allowed *Scared to Death* to be published under his name and with a ghoulish, distorted typeface on the cover. At times in the book, Barker veered away from his main subject to include his work on Aberfan and, after a long disquisition on the ethics of fortune telling and stories of precognition, described the Premonitions Bureau, although he did not spell out his own involvement. Barker also admitted his own fascination and feeling for the occult. 'Much as I dislike it,' he wrote in the introduction, 'I have to admit that I appear to be subject to premonitions, usually non-specific, vague forebodings, but none the less worrying and always followed by some sort of accident or disaster.' In the section on Aberfan, the psychiatrist questioned the uniform progression of time: 'If we are to accept the evidence for precognition from the cases cited here, we are driven to the conclusion that the future does exist here and *now* – at the present moment.'

Scared to Death finished with a rather guileless plea for a fraction of the world's armaments spending to be diverted to fundamental research of this kind. 'When we know more about nature and time and life itself we shall begin to comprehend some of the mysteries of death and what lies beyond for us,' Barker wrote. 'Above all there is a need to banish fear – the fear of the unknown.'

In early 1968, Barker was divided within himself. He was torn between the security and torpor of Shelton and questions that were limitless. There was a realm that he understood, and which bored him. It was almost five years since he and Jane had arrived in Shropshire. He was now the deputy superintendent of Shelton and while his relationship with Littlejohn was problematic, his responsibilities had grown. He ran the hospital's medical advisory committee and stood in for the superintendent at management meetings, discussing replacement boilers, sick pay for the shoemaker and a replacement for the hair perming machine, which had been declared obsolete. Barker was able to further his orthodox medical research at Shelton, which was also novel in its way, but it was a constant struggle. For months, the psychiatrist lobbied to use Hawthorne ward, a deserted isolation unit which had previously housed tuberculosis patients, for his aversion therapy research. He got nowhere. In despair, Barker finally went over Littlejohn's head and wrote to the chairman of the hospital's board for permission. In February, after a day of heavy snow, Barker's request became the subject of a long, stultifying debate among Shelton's senior staff about the manner of his asking (he was reminded not to contact the board directly) and what should be done about the potential infection risk of the ward's old washbasins and toilets. The small-fry politics of the place was maddening.

Outside the hospital, though, Barker's life was as stable and happy as it had ever been. The previous spring, Jane had given birth to their fourth child, Simon. The psychiatrist found himself entranced by the baby's development, his efforts to sit and stand. On a good day, the family house had a buoyant, rambunctious energy. The garden was strewn with trikes and an old pram. There was a rocking horse on the verandah. When the weather was fine, the kids hung upside down on the climbing frame and mucked about on the tennis court. Nigel, who was eight years old, was showing a mechanical bent. He and Barker would walk up the long hill opposite Yockleton, where there was a railway line, and watch the Cambrian Coast Express, a daily steam train, thunder its way to Aberystwyth, on the shores of the Irish Sea. There was some money around. Jane had

her own car, a Mini estate; the children piled in and tumbled about on blankets in the boot. Most mornings, Barker dropped Nigel off at a private school in Shrewsbury before driving on to start his day at the hospital. Having rented Barnfield since 1963, Barker and Jane had recently decided to buy and renovate a house called Bowbells, on a quiet lane near the hospital, closer to town, and settle in the area for good. They hired an architect and spent their weekends picking out furnishings for the family home.

And yet he smouldered. He was forty-three years old. He was stuck in a backwater, attending coroner's inquests of old ladies who fell out of the bath. He railed against 'little men'. He was willing to court a certain notoriety, to put people's noses out of joint, in the course of pursuing questions that he believed were important, or were categorised wrongly as unserious subjects for research. After Barker wrote about the Premonitions Bureau for the *Medical News* the previous year, his article had prompted a stream of letters. Readers complained that the project was unscientific and biased. 'To invent a weird terminology for a series of haphazard guesses and to attempt, unreasonably, to reason with undefined and indefinable data is undignified,' one wrote. Another accused him of being years out of date. Barker did not ignore the letters but replied, apparently calmly, in the newspaper the following week. 'Existing scientific theories must be transformed or disregarded if they cannot explain all the facts,' he wrote. 'Although unpalatable to many, this attitude is clearly essential to all scientific progress.'

If there was a choice between protecting his career and reputation and attempting to prove that our minds could transcend time – and one day, in sufficient numbers and with sufficient data, stop aeroplanes from falling out of the sky – Barker never made it look like one. Leaving these matters aside was incompatible with what he understood to be psychiatry.

On 22 February, the day before *Scared to Death* was published, Barker travelled to Birmingham, where he filmed an interview with the BBC for the evening news. In the living room at Barnfield, his family gathered around the television to watch. In Dagenham, Hencher also happened to see the psychiatrist on the screen, talking of mortal dread. Barker had met Hencher in person the previous month, on a trip to London, and the percipient had repeated his warning: 'I think you're going to have some trouble.' He was sure, like he had been the previous April, that Barker would die at home in Yockleton. Now, watching him on TV, Hencher had the same black feeling as before.

In their letters and telephone calls, Barker had taken to asking Hencher if his travel plans were safe, and then carrying on regardless. He arrived in the capital the following day, to give more interviews. Articles about *Scared to Death* appeared in the *Daily Express*, the *Daily Sketch* and the *Birmingham Evening Mail* on publication day. That evening, Barker appeared on television again. This time, he claimed that lives could have been saved in Aberfan if there had been a properly functioning Premonitions Bureau. 'Despite the climate of public opinion, somebody may have been able to do something to avert the disaster if these people had been able to tell a central body of their premonitions,' he said. The next day, a Friday, Fairley devoted part of his 'World of Science' column in the *Evening Standard* to an idea that Barker had aired in his book: to invite fortune tellers into mental hospitals, to share their intuitions about psychiatric patients.

In the afternoon, Fairley recorded another conversation with Barker for *New Worlds*, a science programme he presented on BBC Radio 4. Afterwards, the psychiatrist was driven out to Elstree Studios, in Hertfordshire, and was paid £50 to talk about his book on *Follow Through*, a late-night magazine show on Associated Television. *Scared to Death* was presented as a scientific study but its themes and case histories were close enough to daily life and family folklore to make it pleasantly unsettling. Barker himself was an ideal foil for journalists – an 'eminent psychiatrist', a 'senior consultant' – a respectable doctor, making unusual claims, and keen to make news. He spent half the week in front of microphones and bright, baking lights. It was exhilarating and exhausting and exactly what he had been told not to do.

*

Barker's television appearance on *Follow Through* was broadcast at 11.45 p.m. on the Saturday. Twenty-four hours later, a woman who became known as Patient 18 woke up in her white-painted cast-iron hospital bed on the first floor of Shelton Hospital and smelled smoke. Patient 18 had been hospitalised at Shelton several times since 1961. Three weeks earlier, she had been admitted to Beech ward, which held the hospital's most agitated female patients, in a 'somewhat disturbed state' but was now considered stable and improving. Forty-two women lived on Beech, day and night, in a long L-shaped set of rooms at the back of the hospital, with a view of the bowling green and the kitchen garden. The ward had its own kitchen, toilets, medical surgery, day room with comfortable chairs and a television, dining room and several coal fires. It had three doors, one of which led to a fire escape, each fitted with a spring lock that closed automatically and could only be opened with a nurse's key.

During the day, five or six nurses were on duty in Beech but during the night there was only one, usually a nurse in her forties named Kathleen Griffiths, who had worked at Shelton for twenty-two years. A junior nurse split her night shift, which lasted twelve hours, between Beech and Chestnut ward, on the floor below. That Sunday night, at around 10 p.m., the last few patients who had been watching television in the day room went to bed. One pinched the end off her cigarette with her fingers. She remembered throwing the stub towards the fireplace. Griffiths turned off the set. A few minutes later, the senior house officer, Dr Varghese Joseph, and the night sister walked through.

At around 11 p.m., Griffiths and the junior nurse on duty, Joyce Lloyd, had a cup of tea in front of a dwindling fire at the opposite end of the ward from the day room. The temperature outside was freezing but the fires in the hospital were allowed to die down from the early evening. Around the nurses, the patients, many of whom were sedated, slept. Beech had two dormitories, connected by an eighty-foot-long corridor that was lined with beds too. Off the corridor, there were also six lockable cells, for the most delusional and unwell. These had old stable-style doors. The ward had been built in 1856. One of the patients had asked for the top half of her door to stay open, for the air.

Patient 18 slept in a bed against the southern wall, not far from a fire alarm, hydrant and hose. When she woke up, Beech ward was lit by the dim nightlights that enabled the nurses to carry out their duties in the dark. If she looked up, she might have noticed that the smoke, which was bitter and tarry, was coming down from the ceiling. Patient 18 rose from her bed and walked to the fire escape. The door was locked. She looked around. There was no sign of Nurse Griffiths, or anyone else on duty. She made her way along the ward corridor, where the smoke was thicker and the beds were almost touching, and found that the door to the stairs was open. Patient 18 went down and found Griffiths in the ward below, talking to two other nurses. She told them about the smoke. There had been no fire training for nurses at Shelton since 1946. Griffiths had spent her entire career at the hospital and had never taken part in a drill. The fire alarm system, which had been installed in 1962, was tested every day at noon. Three large bells and a siren, mounted on the hospital roof, sounded across the grounds. But barely anyone knew how it worked. A few years earlier, Griffiths had come across a small fire on Beech ward and doused a burning chair in the kitchen sink. She told Patient 18 to go back to bed. She also sent her junior colleague, Nurse Lloyd, up the stairs to see what was happening. By the time the two women reached the door of Beech, the smoke was impenetrable. Lloyd saw flames. 'The place was on fire,' she said later.

Griffiths panicked. Rather than go straight up to the ward herself, or raise the alarm, she walked through Chestnut ward and up another set of stairs to enter Beech from the other end. On her way, she passed three fire alarm points but did not break the glass in any of them. When she arrived at Beech ward, another younger colleague, Brenda Cox, from Larch ward, was unravelling a hose. Shelton had 260 telephones on the night of the fire. If someone dialled 111 a dedicated red telephone would ring in the night porter's office. Most of the ward telephones also had 'Emergency' buttons on them but these were confusing to use: if you raised the receiver at the same time, which seemed the natural thing to do, the line became engaged and you couldn't speak to the switchboard. Three nurses tried and failed to report the fire this way. The porter could have freed the lines, if he had called back the right extension and pressed an extra 1. But he didn't. Minutes passed. The fire grew. Smoke spread. Finally, Nurse Scott, in Chestnut, smashed the glass on a fire alarm. This prompted a small metal button to pop out. She pressed it. This was another mistake. The button stopped the alarm again. It was midnight, at least eight minutes after the nurses knew that Beech ward was on fire, before the hospital porter had any idea about the emergency. According to hospital rules, he was not allowed to call the fire brigade himself. He had to call Shelton's deputy fire officer first, who was asleep at home.

The first firefighters arrived on the grass below Beech at thirteen minutes past midnight. According to the coroner, most of the women died of carbon monoxide poisoning as the smoke enveloped their beds, a process that probably took about six minutes. Before the fire crews arrived, a student nurse named Dennis Lewis had saved several patients. He crawled along the floor, where the air was still breathable, tugging sleeping bodies out of their beds. A policeman scaled some builders' scaffolding and got five women out of a

197

dormitory window. Firemen with breathing apparatus broke down the doors of the locked cells and found five out of the six patients alive. The woman who had asked for her stable door to stay open was dead, along with the eight women who had been sleeping in the corridor outside. A nurse who visited the ward several weeks later found huddled shapes burned into the bathroom tiles.

The fire was under control by 2 a.m. The hospital set up a temporary mortuary. Twenty-four women lost their lives. It was the worst fire in a British hospital since 1903, which also occurred in a Victorian asylum. More than a hundred patients were evacuated from their wards across the building. Nurses set up makeshift beds in the hospital's main hall and sedated those who were disoriented and bothering the rest. Shelton's chief nurse, Arthur Morris, was picked up at home by the police and put in charge. When a firefighter carried out the body of a woman who became known as Patient 27, a letter fell out of her bedclothes on to the floor. It was addressed to her father and finished: 'I hope the Nurses and the Girls go on a blazing hot fire, Dad.'

Rumours flew. Several patients claimed to have started the fire. Patient 23, a chronic patient with schizophrenia, who had been at Shelton for twenty-nine years and who slept at the top of the stairs leading to Chestnut ward, swore that it was her. 'I fetched the wax,' she said, claiming to have set alight a seven-pound jam jar full of furniture polish that she found in a store cupboard. She grinned at the memory. A nurse who questioned Patient 23 believed her. She was one of David Enoch's patients. At the public inquiry later that year, Enoch testified that she exhibited echolalia, a symptom named after the Greek nymph Echo, who was punished by Hera simply to repeat everything that she heard.

Overnight, Shelton became a symbol of everything that was wrong in Britain's retrograde mental hospitals. Newspapers printed photographs of firefighters next to burned-out, jumbled beds. The hospital's southern walls were shown gothic and fire-blackened. In the House of Commons the following afternoon, the Conservative MP for Shrewsbury, a former fighter pilot named Sir John Langford-Holt, drew attention to Shelton's age. 'This hospital was constructed fourteen years before the Indian Mutiny,' he said, 'and this probably goes right to the heart of the problem.' Press accounts of the fire described Beech ward's locked doors and raised the question of whether the patients had been left unattended. Shrewsbury dignitaries told the press that the hospital should have been condemned years ago. Lewis Motley, a former chairman of Shelton's management committee, described the place as a snakepit. 'You could smell it a mile off,' he said.

'An ancient monument' was the headline in the local newspaper, the *Shropshire Star*. 'What has always been a mystery to us is how the depleted, harassed and overworked medical and nursing staffs have been able to care successfully for anyone in such surroundings,' ran an editorial. 'Shelton Mental Hospital still looks like one of the old institutions and is often locally regarded as one.'

199

As a senior doctor at the hospital, Barker felt a deep sense of shame. Two days after the disaster, he wrote to Barbara Robb, author of *Sans Everything*, and urged her to use the material he had gathered about the hospital's static, age-ing population. His research, he wrote, 'takes on new and pressing meaning in the light of the tragedy that took place at this hospital this weekend'. He pointed out that many of the victims were elderly, chronic patients and offered to supply Robb with further confidential information from Shelton. 'I must admit I detest locked wards and closed doors,' Barker wrote. 'They are an anachronism.' On 8 March, twelve days after the disaster, the *Nursing Mirror* published an article by Barker and Mabel Miller about their survey and illus-trated it with a photograph of Beech ward, gutted and with its ceiling black and blistered by the fire. The two doctors' study, the newspaper claimed, revealed 'conditions there which, unfortunately, are typical of many such hospitals in this country'.

Barker had submitted the text before the fire but its pub-lication made him deeply unpopular among the hospital's traumatised staff. A discussion of the matter at the monthly management meeting was adjourned so everyone could read the article and digest the full extent of Barker's insensitivity. Littlejohn convened a gathering of the nursing staff, who felt most abused by the article, so they could reprimand Barker in person. The psychiatrist asked to tape record the meeting. Littlejohn refused. 'We are now having a terrible time at the hands of the critics,' Barker wrote to Robb.

His own position at Shelton seemed secure, but he was sure that Littlejohn and other members of the hospital hierarchy were trying to make his life as difficult as possible. *Scared to Death* was due to be published in America later in the spring and Barker was planning a lecture tour to talk about the book, as well as his research into aversion therapy and the Premonitions Bureau. Miller was supposed to come too but when she asked for study leave to accompany Barker, the hospital refused. Miller asked why. The hospital secretary replied that he was unable to say.

The dispute infuriated Barker. He did not like to say that he was afraid to travel alone. He complained instead that the place was so small-minded. If the fire proved anything, it was that hospitals like Shelton could not be reformed quickly enough. But the psychiatrist's abrasive manner didn't help him. Nor did his hunger for the limelight. Barker's dalliance with the occult and his arguments for the Premonitions Bureau were public knowledge now. *Private Eye*, the London-based satirical magazine, noted acidly that Barker's brainchild had failed to foresee a disaster at his own place of work. 'Of course, every scheme has its teething troubles,' the magazine reported, 'but it is a pity that fate should without warning turn its hand to Shelton Hospital, Shrewsbury.'

*

The bureau had been open for fifteen months. By the spring of 1968, Barker had a collection of 723 predictions from the public. On consecutive days in March, Fairley wrote up the findings from the first year of the Premonitions Bureau in the *Evening Standard*. Based on his scoring system, Fairley calculated that eighteen of the warnings received during 1967 had come true so far – a hit rate of slightly more than three per cent. It was very low, but not nothing. And most of the predictions were still that; the future was wide open. 'So little is known of the role of time in this phenomenon that it is impossible to say they will never come true,' Fairley wrote. He announced that the *Standard* would run the experiment for another year 'to amass more material and observe certain people more closely'.

The bureau's strike rate was more impressive if you concentrated on the visions of Miss Middleton and Alan Hencher, who, among their many warnings, had contributed twelve out of the eighteen apparently successful premonitions. 'These two, if the evidence is accepted, appear to act as "human seismometers", who get early warning of disaster,' Fairley reported. Hencher claimed to have foreseen the Cyprus and Stockport air crashes, Miss Middleton had the death of Vladimir Komarov, and both seers had given convincing warnings of the Hither Green crash. A few days after Christmas, Miss Middleton had also envisaged an unusual collision involving a truck carrying an 'exceptionally heavy load'. Seven days later, on 6 January 1968, a low-loader carrying an electrical transformer, which weighed 120 tons, was hit by an express train on its way from Manchester to London on a level crossing at Hixon in Staffordshire. Eleven people were killed.

In Fairley's second article about the bureau, he described how the music teacher and the telephone operator perceived their visions. 'I see a picture projected in front of my eyes,' Miss Middleton said. Often she saw a single object – a building, a train or a car. Words flashed, as if in neon. A day or two later, she would have the same vision again, but more details would be revealed. Hencher's premonitions were always accompanied by pain. The back of his skull, where he suffered his head injury, ached. It felt like a migraine. 'Sometimes I see things in black and white, sometimes in colour,' Hencher told Fairley. 'I never get pleasant premonitions.' When he sent in his warnings, Hencher said, the pain went away. When the disaster occurred, the words 'It has happened' occurred blankly in his mind, either as words or a voice that only he could hear. 'Imagination? Maybe,' Fairley wrote. 'Mr Hencher is simply bewildered.'

The percipients were photographed for the *Standard*. Miss Middleton smiled gamely. Hencher, wearing a V-necked sweater, held a pair of secateurs to a wintry tree. The attention was stimulating. But it also caused problems. Early in the experiment, Hencher had complained to Barker about how paying attention to all of his visions was affecting his mental health. This feeling had festered. Now Miss Middleton was beginning to feel exploited too.

After they were identified as the experiment's stars, Hencher and Miss Middleton, who were each isolated in their own way, decided to join forces. The bureau became cognisant. On 8 April, in a letter to Fairley timed at 6.22 a.m., Hencher reported his latest premonition: an aircraft, seventy-four passengers, tipped on its side, 'Finland comes into this, don't know why.' In the same letter, he raised a number of grievances and made it clear that he and Miss Middleton had been comparing notes. 'If you would wish all and anything to be recorded, then we will do so but this is going to take up a lot of time,' Hencher wrote. Jennifer Preston was on maternity leave, giving birth to her third son. Hencher said that he hoped 'us Strange Bods' were not alarming Fairley's temporary assistant in her absence. 'I'd hate to give her night-mares,' he wrote.

In the following days, Hencher sent Fairley two further letters, of increasing intensity, making it clear that things couldn't go on as before. He and Miss Middleton wanted reports of all their premonitions to be returned. They were collaborating on a book. 'It is going to be rather outspoken on certain points, and will probably annoy various people,' he warned. 'Do those who record our predictions really understand what it entails to make predictions?' Then, in a strained passage, the seer tried to make Fairley understand what it was like to interrogate the partial, frightening patterns of your mind, come off a night shift at the Post Office switchboard into the London dawn, sleep in your council house in Dagenham, and then see your visions splashed all over the evening papers.

We have to undergo the torment of knowing that whatever we say, however long we pray, that when we receive, we have the problem of deciding whether we should tell what we have received – because if we dont and it happens we cannot be believed, and so suffer equally by the knowledge that we are doing something that could be of public use, and by reason of the partial publicity, the here and there perhaps, if it is sensational enough, that this very public must start to see us as cranks, so creating even greater stress mentally to the nervous system.

Hencher's anguish was real.

if we dont and it happens we cannot be believed

205

When every thought can be a sign, when psychiatrists and newspapermen are hanging on your every strange notion, which do you choose to record and which do you wave away? Do you even get to choose?

whatever we say, however long we pray

The responsibility felt overpowering. Hencher dreaded the pain that came with foreknowledge. He wanted it to be worth it. So did Miss Middleton. They wanted recognition and a little bit of looking after. Some money wouldn't hurt. 'We have reached the point of no return,' Hencher wrote. 'We have become known, this cannot be undone by

A probe by a special Evening Standard bureau
PREMONITIONS
The Londoners who believe they saw disaster in advance

This was the news headline in the Evening Standard yesterday. Now PETER FAIRLEY gives more details recorded by the Evening Standard Premonitions Bureau which records premonitions before the forecast events happen . . .

Did Mr. Hencher forecast th

turning the clock back.' He was suspicious of what Fairley and the *Evening Standard* were getting out of the whole experiment.

The seers were never as respectful towards Fairley as they were towards Barker. Fairley was never quite as taken by them, either. He waited five days before replying to Hencher's ultimatum. 'While I understand that to both Miss Middleton and yourself the experiences you undergo are disturbing and intriguing,' Fairley wrote, 'I would urge you to resist the temptation to exaggerate them and claim for yourselves a "power" which you may not, in fact, possess.' The science writer did not have so much riding on the Premonitions Bureau. He was always on the move, looking for the next big reveal. When he

EVENING STANDARD, TUESDAY, MARCH 12, 1968—7

ALAN HENCHER : I NEVER GET PLEASANT PREMONITIONS

lither Green rail disaster?

had lost his sight, in 1966, he wrote about his recovery and his intention to pursue a gentler kind of life. 'There is no such creature as Superman. He is a myth,' Fairley wrote. But even as he conceded this, Fairley ended up describing the person that he always would be: 'I realise that no man, no matter how easily he can shrug off minor infections, run for a bus, outsmart his rivals or achieve his personal targets, can assume he is indestructible.'

Fairley was more likely to hail a taxi than run for the bus. But he continued to travel, to hustle the competition, and to swipe restaurant bills that other people paid, in order to pad out his expenses claims. He talked his way into jazz clubs. He had affairs. By the end of the sixties, Fairley's contacts in the US and Soviet space establishments were paying off in style. At the time of the Hither Green crash the previous November, which had been so significant for the others involved, Fairley had been in Florida, reporting on the next stage of the Apollo programme.

On the foggy, grey morning after the accident, when Barker was giving interviews in London, Fairley was standing in the sunshine in front of Launch Complex 39a at the Kennedy Space Center in Florida, where Saturn V, the giant rocket that would carry Apollo 11 to the moon, stood attached to its orange iron tower. The rocket was sixty feet taller than the Statue of Liberty. An ITN cameraman was positioned a quarter of a mile away with a huge zoom lens focused on Fairley's face.

'They are calling this the Big Shot,' Fairley said.

On the word 'big', the camera panned out rapidly, filling the frame with the Saturn V and reducing Fairley to a pixel in the landscape.

'It said it all in one,' Fairley recalled in delight. He loved television for moments such as these. He dressed up in space-suits or NASA overalls at the merest excuse. When Saturn V blasted off, three days later, Fairley was in the press gallery, at a theoretically safe distance of three and a half miles. It was 7 a.m., dead calm, with the dawn washing away last night's clouds. Children sat on car roofs. Men looked up to the sky. When the rocket's five F-1 engines fired, the launch pad became a skirt of flames. For an instant, Fairley thought something had gone wrong. Then the giant column, weighing almost three thousand tons, defied sense and rose into the air. Fairley felt his ribs shake. People cheered in disbelief at the might of it all. He heard himself shouting, 'Go! Go! Go!' The glass walls of Walter Cronkite's commentary box, for CBS News, broke apart.

Space was taking over Fairley's professional life. In the spring of 1968, he decided to leave the *Evening Standard*. British independent television was growing and consolidating. The *TV Times*, a magazine which had appeared in regional editions, was about to become a national title. Fairley was offered the position of science editor for both the magazine and ITN – a dual presence, in print and on screen. He gave his notice to Charles Wintour and saw out his last few months at the newspaper, going for lunch with his new colleagues on Tottenham Court Road and plotting the next stage of his career. The first national edition of the *TV Times*, published later that year, showed Saturn V blasting off and a freckled seven-year-old boy in a spacesuit (the son of one of Fairley's NASA friends, who happened to be British) with the headline: 'Introducing . . . the first Englishman to set foot on the moon', a typical piece of Fairley showmanship.

He didn't give up the Premonitions Bureau. Fairley planned to take its records with him to the *TV Times*. He asked Preston to come with him. He wanted to see where the experiment went; it might make good television. When he pondered how precognition might work, Fairley's mind often drifted to space and its abstractions. He was fascinated by Lagrange points. These are places, named for Joseph-Louis Lagrange, a French-Italian mathematician, where the gravitational fields of celestial bodies cancel each other out and an object can hang motionless, in theory, forever.

There is a Lagrange point on the way to the moon, where the pull of the Earth gives way to the pull of her satellite. In an interview with the BBC in 1977 about his interest in parapsychology, Fairley wondered aloud if an astronaut could leave a thought at this point and it could then slip into the mind of another astronaut, following months behind. When Fairley was challenged by the interviewer that this sounded like science fiction, he agreed. 'One of the deep worries is that it is not studied properly,' he said.

The conversation moved to the question of premonitions. 'It is a fascinating subject,' Fairley said. 'So many people know that it happens . . .'

'*Believe* that it happens,' the interviewer corrected him.

'I say that I know that these things happen,' Fairley said. 'Because they happen to me.'

Still, there was a line somewhere. And Fairley, a man about town, was wary of the suburban seers – the slightly embarrassing figures in David Frost's green room – in a way that Barker never was. He was also convinced that premonitions took place in an unthinking realm. His gambling streaks on horse races went well, until he found himself considering his choices. 'When you start to think about these things, you might as well forget it,' Fairley told the BBC. 'This is why one has to be terribly careful about these people who make a profession or even an amateur play to predicting the future. The moment they appear to be using some kind of thought of projecting the future, I am deeply suspicious.' On 17 April, Fairley forwarded his recent correspondence with Hencher to Barker, suggesting that it was time to break off contact with the bureau's most successful predictors: 'It appears from these, that neither he nor Miss Middleton can be any longer of use from a scientific standpoint!'

Barker wasn't so sure. 'Boastfulness is unfortunately a common concomitant in my experience with clairvoyants,' he replied. The psychiatrist was fond of Miss Middleton. He found her 'a pleasing personality'. He agreed that Hencher was paranoid but, as ever, he found the seer's disquiet a potential object of study, instead of a reason to end the relationship. Barker suggested that they continue to collect the visions of their best contributors for the rest of 1968 and try and grow the pool of percipients, rather than cull them. 'I have always thought that there must be many others with similar abilities that we have not as yet communicated with presumably because they do not know of the existence of the premonitions bureau,' he wrote.

If anything, the tone of Barker's letter was diffident, a little detached. 'I would not like to say at this stage . . .' He played for time. He, too, was also occupied by other things. After the thrill and the glamour of the publication of *Scared to Death*, the fire on Beech ward had come close to shattering Shelton as an institution and Barker's mission as a doctor there. And yet there he was, pinned in the Shropshire countryside, under trees that were overgrown and blocked out the light. Barker passed the ruined ward every day. He longed to be elsewhere. If the future did exist here and now – if Barker's future was already present – it was out of his reach. It hovered outside the hospital walls.

*

Early one afternoon in May, a BOAC airliner, with a grey belly and a navy blue trim, swung west, reached its cruising altitude over the Irish Sea and settled on its course for New York. The crossing was smooth. The air stewardesses wore white gloves. Barker, who was a nervous flyer, sat in a window seat. The psychiatrist had left Yockleton early that morning with Miller, his research assistant, with their luggage and slides for a three-week lecture tour of the US. Miller, who had been denied study leave by the hospital, had been forced to take the trip as a holiday. It was Barker's longest separation from Jane since the children were born. As the VC10 skirted the coast of Northern Ireland, the clouds allowed a glimpse of the seaside towns of Coleraine and Portrush, where he had spent time as a teenager when his father was posted to Belfast during the war. His thoughts unspooled, backwards and forwards across time.

The official, scientific purpose of the trip was for Barker and Miller to visit universities and state psychiatric hospitals to talk about aversion therapy, which had made an impact in the American press. But there was also, as always, the matter of the supernatural. *Scared to Death* had gone straight to paperback in the US, where there was no attempt by Barker's American publishers to present his work in a sober or scientific light. 'Thoughts can kill!' ran the cover line. Below the title, a woman in a white dress, the hem bunched in her hand, ran from an apparition. A bulging eye stared through a rip in the image.

There were outlets, both fringe and mainstream, for Barker's ideas in America. There were ESP research projects that were much more dynamic and better funded than anything taking place in Britain at the time. Barker's itinerary included visits to the Maimonides Medical Center in Brooklyn, where Montague Ullman, a psychiatrist, had been running dream and telepathy experiments since 1962.

Thoughts can kill! A medical doctor's amazing,
fully documented case histories of deaths caused by
strange and terrifying psychic powers

SCARED TO DEATH

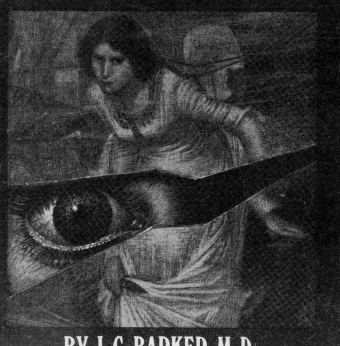

BY J. C. BARKER, M.D.

Barker was also hoping to address the American Society for Psychical Research, which was undergoing a revival thanks to donations from Chester Carlson, the inventor of the Xerox process and a believer in life after death.

The tour came at a propitious moment. It was a chance to break the monotony of Shelton, perhaps permanently. It was a way to step outside himself. But there was also something unreal and overwhelming about the adventure. Barker referred to his journey as the 'trip of a lifetime'. He was sure that nothing like this would happen to him again and there was, mixed up in the excitement, a fateful air to his departure. Just after 3 p.m., over the Atlantic, Barker wrote out two letters on BOAC headed paper with his fountain pen. The first was to his parents. 'Things are still awful at Shelton,' he wrote. He had argued again with Littlejohn – 'a vile character' – the day before. The second, to Jane, was softer and more wistful. The family had returned to Woolacombe over Easter and Barker replayed scenes from the holiday in his mind. The new house, Bowbells, was coming along. He dreaded the thought of being trapped at Shelton but he sought some consolation in the stability that he was now able to provide. 'At last you can say that that will be something I have done for you (after Herrison etc),' he wrote to Jane. He told her not to worry about spending money while he was gone. 'What the hell does money matter now?'

The tour, when it began, was gruelling. On the night that Barker arrived in New York, he gave radio interviews about both *Scared to Death* and aversion therapy and reached his room at the Manhattan Hotel, on Eighth Avenue, at 3 a.m., a little over twenty-four hours after leaving Barnfield. After a day to recover, he and Miller headed to Philadelphia to give a lecture and then flew up to Rochester, on the shore of Lake Ontario, to present their work at the university hospital. Over the next ten days, Barker and Miller criss-crossed the eastern seaboard, visiting medical schools and mental hospitals in a rented Dodge Monaco. They saw Niagara Falls in the rain.

The distances stunned Barker. There were days when he looked at the map and it seemed like they had made no progress at all. Some places he loved. The beauty of Vermont reminded him of the Wye Valley, on the Welsh borders, south of Shrewsbury. He bought a clock to add to his collection. On Cape Cod, Miller photographed Barker resting on a rock, with the sea lapping quietly behind. He wore a brown long-sleeved polo shirt. His black, greying hair flopped untidily over a receding hairline. He had a contained, quietly belligerent energy. As they drove, Barker and Miller picked up on the fractiousness of America that spring. The Vietnam War was at its height and had become patently unwinnable. Martin Luther King had been assassinated on his hotel balcony in Memphis just over a month before. One day, probably near Hyannis Port, on their way to Boston, the two doctors caught sight of Bobby Kennedy, who was on his late, electrifying campaign to seize the Democratic nomination for the 1968 presidential election. The big cities made Barker uneasy. 'We like America, but it is a sick society,' he wrote to Jane. Philadelphia was horrible. The addicts and the drifters in New York scared him. He found the Manhattan skyline extraordinarily phallic; he couldn't stop staring at it.

In Boston, the city of Miss Middleton's birth, Barker and Miller presented their aversion therapy research to five hundred doctors at the annual meeting of the American Psychiatric Association, at the Sheraton Hotel. Erwin Stengler, a student of Freud's and a world authority on suicide, was in the audience. There was laughter during the presentation, which unnerved Barker. He couldn't tell whether it was a good sign. 'I don't think I am going down here as well as I would have liked,' he wrote to Jane. Most of the hospitals depressed him. 'We lectured here at the state mental hospital,' he wrote from Ancora, New Jersey. 'Which is bloody awful.' Barker exhibited a certain disdain towards the people and the places that he visited, while hoping for a job offer and research opportunities at the same time. He recognised his own ambivalence as part of the problem. 'I don't think I like American psychiatry all that much – but then I don't like British either!' he wrote. The schedule wore him down. Miller developed a stomach bug. But there was no shortage of opportunities to speak and the two of them pressed on. A fortnight into the tour, Barker and Miller returned to New York, where they crammed in extra meetings and seminars. They gave a two-and-a-half-hour presentation at a meeting of behavioural therapists on a Saturday morning. For several days, Barker lectured on aversion therapy in the morning and *Scared to Death* and precognition in the afternoon.

At the Maimonides Medical Center, the parapsychologists from Ullman's Dream Laboratory came to hear about the Premonitions Bureau. Among them was Robert Nelson, a thirty-year-old volunteer at the lab, who worked in the circulation department of the *New York Times*. Nelson, who was blond, blue-eyed and from Ohio, was part of the folk scene in Greenwich Village. He had a twin brother, William, who was a medium. The following month, Nelson set up an American version of the experiment, called the Central Premonitions Registry, which he ran from PO Box 482 in Times Square station.

On 26 May, Miller returned to England, taking most of the lecture equipment with her. Barker stayed on to spend an extra week in California. On the flight from New York, he watched a cold front roll in over the Rockies. Turbulence began to shake the plane as he wrote a seven-page letter to Jane. 'I must say I am very frightened of doing it alone,' Barker confided. 'However, it would be a pity to miss the opportunity at this stage – probably the only one I'll ever get.' The storm worsened as Barker's letter progressed. 'It's rocking like a palm tree!' He added in a PS: 'Scared to Death!'

In Los Angeles, the psychiatrist's days and nights were emptier. He took a bus tour and inspected the hand and footprints of the movie stars outside the Chinese Theatre on Hollywood Boulevard. Barker found that he had the same size feet as Clark Gable. The city was endless, and sweltering. He watched the hippies on the Sunset Strip. He sent his family a postcard of the Harbor Freeway. 'I have never seen so many cars.' One day, a helicopter took him to San Bernardino, to speak at the Patton State Hospital, and Barker was finally offered a job, working at California's psychiatric prison hospital on a salary of $17,000 a year, more than he earned at Shelton. 'I doubt if you would like it,' he wrote to Jane. He struggled to sleep in his downtown hotel.

*

Each day, the news was full of Bobby Kennedy. Voting in the California primary was less than a week away. The campaign headquarters was based at the Ambassador Hotel, not far from where Barker was staying. Two days before the psychiatrist arrived in the city, Kennedy had appeared at a primetime televised gala at the Los Angeles sports arena, in Exposition Park. The evening was hosted by Kennedy's friend Andy Williams, who sang 'Moon River'. The Byrds played a Bob Dylan song. Jerry Lewis did a skit. So did Gene Kelly. Raquel Welch wore a floor-length white dress. Ethel Kennedy, Bobby's wife, was there, pregnant with the couple's eleventh child.

It was a Friday night and Kennedy, who was forty-two, had been campaigning in Oregon, a thousand miles to the north. He arrived at the gala late, after his fifth plane ride of the day.

In his breathless eighty-day run for the presidency, Kennedy harnessed the glamour and wealth of his family's political machine to an emotional, morally charged appeal to end the war in Vietnam and address poverty and racism in America. He let crowds take his shoes. His hand was pummelled and shaken until it bled. He drank ginger ale to keep going. Onstage in LA, he was introduced by Shirley MacLaine, the star of *The Apartment*. Kennedy was slender in black tie. He told a few jokes about Ronald Reagan, the governor of California, and Sam Yorty, the mayor of Los Angeles, who hated him, before he shifted to the campaign's searching tenor, in which Kennedy would quote alternately from John Donne, Albert Camus and his favourite piece of ancient graffiti, which was said to have been etched into a brick on the pyramids: 'And no one was angry enough to speak out.'

Miss Middleton was certain that Kennedy would be killed. She claimed that her Massachusetts upbringing gave her special insight into the family. On 11 March she had written to Barker, warning of an assassination. Four days later, she wrote again: 'The word assassination continues. I cannot disconnect it from Robert Kennedy. It may be history will repeat itself.' Later that month, Miss Middleton told an American journalist that two men weren't safe: Charles de Gaulle and Bobby Kennedy. She repeated her warnings about Kennedy throughout April.

She wasn't the only one. Death threats arrived regularly at the campaign office. Each week Kennedy's press secretary, Frank Mankiewicz, was given images of potential shooters

by the FBI, and he scanned for their faces in the crowds. Kennedy himself was fatalistic. He had a single bodyguard. 'Living every day is like Russian roulette,' he told Jack Newfield, a reporter and later biographer. 'You've just got to give yourself to the people and to trust them, and from then on it's just that good old bitch, luck . . . I am pretty sure there will be an attempt on my life sooner or later. Not so much for political reasons. I don't believe that. Plain nuttiness, that's all. There's plenty of that around.'

After his brother's death in 1963, Kennedy experienced an extreme form of survivor's guilt, what Milton, in *Paradise Lost*, called overliving. He sought consolation in *The Greek Way*, a guide to ancient Greek thought by Edith Hamilton, a schoolmistress from Fort Wayne, Indiana. Hamilton was not a conventional scholar but a talented populariser, with a feel for language and an instinct for the redemptive possibilities of suffering. It was Hamilton's free translations of Aeschylus, and hybridised classical sayings all of her own, that peppered Kennedy's speeches in 1968.

'Men are not made for safe havens,' Hamilton wrote of Aeschylus in a passage that Kennedy underlined. 'The fullness of life is in the hazards of life. And, at the worst, there is that in us which can turn defeat into victory.' Part of tragedy is the certainty by which it proceeds and how we interrogate our choices as it does so. 'My God, it might have been me,' Kennedy said, after the shooting of King that spring. That night, the candidate spoke beautifully and sadly of his brother's death to a shocked crowd in a Black neighbourhood of Indianapolis. He soothed their pain by misquoting Hamilton's own botched translation of Aeschylus.

*

Greek philosophers realised that a person's fate is inseparable from who they are. What is going to happen becomes something that we choose to do. Our actions are both an expression of our distinctive character, our *ethos*, and a divine force that plays on us, our *daimon*. Socrates had a vocal *daimon*. When he was put on trial, in 399 BC, one of the charges that he faced was of introducing new gods. Socrates had caused unease in Athens by referring openly to the clairvoyant spirit that guided him. 'This sign I have had ever since I was a child,' he told his jurors, five hundred fellow citizens who had gathered in the agora to judge him, according to Plato. 'The sign is a voice which comes to me and always forbids me to do something which I am going to do, but never commands me to do anything.' Socrates's commitment to his oracle was absolute. During his trial, he could have asked for mercy, or gone into exile, or paid a fine, at any stage. He was seventy years old. He had no shortage of connections. But he did the opposite. He provoked his prosecutors. He lectured them. He demanded reward instead of punishment. The warning voice never came. Even when Socrates was condemned to death, the *daimon* did not tell him to change course. His friends came to his cell to rescue him; he sent them away. Socrates concluded that there was nothing to fear. 'A wonderful thing has happened to me,' he said. 'This thing which might be thought, and is generally considered, the greatest of evils has come upon me; but the divine sign did not oppose me.' If there is no disaster, there is no premonition to precede it. Socrates raised the hemlock to his own lips. 'That which has happened to me is doubtless a good thing,' he reasoned, 'and those of us who think death is an evil must be mistaken.'

*

Barker spent the final days of the tour in San Francisco. The Pacific lured him, but he was afraid to surf. He felt lethargic and alone. 'At the moment, I just feel like curling up and dying, I so want to get home to you all.' At the same time, the psychiatrist dreaded the resumption of

his former life. 'I expect trouble on my return to Shelton,' he wrote. 'I wonder what future obstacles will be placed in the way.' Barker reached Shrewsbury on 1 June, which was a Saturday. The following Monday was the last day of campaigning in the California primary. Late in the morning, while he was being jostled and grabbed by a crowd in Chinatown, in San Francisco, a firecracker exploded near Bobby Kennedy. There was a string of loud bangs. His entourage cowered. He carried on shaking hands. The next day, Miss Middleton was frantic. 'Another assassination and again in America,' she wrote to Barker. She called the Premonitions Bureau three times on 4 June, warning that a killing was imminent. That afternoon, at the beach in Los Angeles, Kennedy's twelve-year-old son, David, got into trouble in the undertow, and he dived in to save the boy. He was shot in the head shortly after midnight, as he cut through the kitchen of the Ambassador Hotel, minutes after declaring victory in the California primary. 'Everything's going to be okay,' Kennedy whispered, as he lay dying on the floor. Barker described it as Miss Middleton's best prediction yet. 'You were insistent,' he wrote.

*

The dust fell in many colours. People woke up and saw it: streaks of orange on the bus windows, a smattering of yellow on the paving stones. Summer leaves were freckled white. Where it was dry, the dust had the consistency of make-up; so fine you could barely sense it on your fingers. Where it mingled with rain, the gutters ran red. It was a Monday morning, 1 July 1968. The air felt ill. Across the east of England, a heat wave was burning out. London overheated. Tower Bridge jammed open. Members of Parliament asked permission to take off their jackets, and were refused. From the north and the west, cold wet air moved in from the Atlantic. Hanging in the sky was ten million tons of Saharan sand which had billowed up from the Ahaggar mountains in Algeria.

Everything collided, more or less, over Shrewsbury. Meteorologists described a squall line that ran up the Welsh border, and along that line people experienced the strangest morning of British weather since 1755. Ice fell and cluttered the roads. Greenhouses smeared by coloured rain were then smashed by hailstones the size of golf balls. The town of Welshpool was cut off by floodwater. An old man who was chopping wood in his shed during the downpour was swept into a river, along with his shed, and drowned. Students were asked to lift their feet while their exam halls flooded. In Shrewsbury, lightning ripped a hole in the roof of a house and scorched the living-room carpet, while a seventy-three-year-old woman watched television alone. The storm turned the sky green and black at noon. In Stockton-on-Tees, drivers turned on their headlights, manhole covers were pushed upwards by the rain, and people knelt in the streets and prayed. Aberfan flooded again.

Amid thunder, the public inquiry into the fire at Shelton

began. The hearings were held at Shrewsbury's Assize Court, which was part of the county council's brand new head-quarters on the outskirts of town. The modernist Shire Hall, a world away from the gothic corridors of the hospital, had been opened by the Queen the previous spring. Rain battered the narrow rectangular windows. Barker sat in the public gallery. Over the next two weeks, the inquiry took testimony from forty witnesses. The psychiatrist listened while the nurses contradicted one another and gave muddled explanations for what happened that night. The hospital's resident fireman, Joseph Wade, explained that for eleven years he had been 'just a glorified labourer' and forbidden to talk to the female staff. Enoch appeared at the inquiry and presented his diag-nosis of echolalia. Barker sensed a cover-up. 'The conflicting evidence is quite horrifying,' he wrote to Barbara Robb, 'and I am sure is giving this hospital a very bad name.'

On 11 July, Barker turned forty-four. It was his father's seventy-seventh birthday the following day. The men exchanged cheques. Jane sent Charlie some tobacco. Since he had returned from the US, Barker had found it hard to get going. He was still lobbying for a proper space at the hospital to conduct his aversion therapy research. As he predicted, Miller was punished for accompanying him on the lecture tour. She had her pay suspended and then docked by three days for going over her holiday allowance. Barker worried that she would resign, leaving him more isolated than before. 'I am ashamed to be associated with it,' he wrote of Shelton in his birthday letter to Charlie. He described it as a cess pit. The children were well. The long summer holidays were about to begin. Bowbells, finally, was nearing completion. He and Jane were planning the move from Yockleton the following month. But Barker was afflicted by a sense of hopelessness, of emptiness, about his own future. 'It seems so futile to have to live here,' he wrote.

Towards the end of the month, Barker began to suffer headaches, which worsened. The pain was often intolerable. He was admitted to Copthorne Hospital in Shrewsbury, where he continued to work and write letters from his bed. 'He was not the kind of person to "give in" – and always wanted to get on with "the job",' Jane wrote later.

On the night of 27 July, Miss Middleton had another dream, which she interpreted as a further warning to Barker. She was staying in a boarding house by the sea. Her dead parents were with her. 'For a brief period, we were happy and had tea,' she recalled. Then her mother rose and climbed into a black car, pushing Miss Middleton away. She chased after the car briefly, and afterwards understood the dream to indicate the passing of someone close to her. When she woke up, Miss Middleton felt like she was in a trance: dulled and detached from the world. At lunchtime, she posted a note to the Premonitions Bureau: 'THIS MAY MEAN A DEATH.'

*

In the early sixties, a French historian, Philippe Ariès, who had already written a radical book about childhood, became deeply occupied with the manner of our dying. Ariès never held a university position. For thirty-five years, he ran the document department of a research institute devoted to tropical fruits. Occasionally, he was mocked by more credentialed academics as 'the banana seller'. He read Latin on the train to work. Ariès wondered whether France's funerary customs – the pious walk to the graveyard; the veneration of tombs – were age-old or more recently invented. He investigated the digging of Paris's great modern cemeteries, in the late eighteenth century. During his research, Ariès glimpsed an earlier, tantalising world of reused graves and mingled bones, where men and women responded differently to the end of their lives. Once he began studying old rites and the verses of medieval *danses macabres*, Ariès found that he could not stop. With his wife, Primerose, he began to visit the national archives, where they spent three years of weekends reading old wills, from the nineteenth century back to the sixteenth. He gave himself to the story of death. 'There was no turning back!' Ariès recalled. 'I had lost all freedom; from now on I was totally caught up in a search that constantly expanded.' Ariès came to the conclusion that, over the course of a thousand years, death had become increasingly private, to the point of invisibility. In the process, it had grown wild. In the early Middle Ages, death had been more commonplace, a simpler and more collective act. 'We all die.' It was the sign of a good life to know that the end was at hand. A bell would ring by itself. A man would hear three knocks on the floor of his room. An inscription from

1151, in Toulouse, told how the sacristan of Saint-Paul-de-Norbonne 'saw death standing beside him', made his will, prayed and died. In Arthurian legend, King Ban watched his castle burn, fell off his horse and looked up to the sky, beseeching, 'Oh Lord God . . . help me, for I see and I know that my end has come.' *I see and I know*. Ariès italicised the words. Gawain, Arthur's nephew, is asked: 'Ah, good lord, think you then so soon to die?' He replies: 'Yes. I tell you that I shall not live two days.'

Neither his doctor nor his friends nor the priests (the latter are absent or forgotten) know as much about it as he. Only the dying man can tell how much time he has left.

Ariès was fascinated whenever a fragment of the old ways survived. In 1959, almost twenty years before Ariès published *The Hour of Our Death*, a retired shoeshine named Malete Hanzakos began to prepare for the end of his life in Bucyrus, in northern Ohio. Hanzakos, who was known as Mike, migrated from Sparta, in Greece, to New York City after the First World War. He settled in Bucyrus in the thirties, where he cleaned shoes, grew vegetables and drove his panelled truck through red lights, stopping when they turned green. He never married or spoke much English. At the age of seventy-seven he suffered no more than the usual aches and pains. In the last year of his life, Hanzakos chose a cemetery plot, had a headstone engraved (except the final date), tended his grave, ordered flowers for his funeral (tied in a ribbon of white and blue, the national colours of Greece) and wrote an obituary for the local newspaper, which refused to print it while he was still alive. He could have been reading from *The Rule and Exercises of Holy Dying*, written by Jeremy Taylor, an Anglican vicar, in 1651: 'Death hath come so near to you as to fetch a portion of your very heart; and now you cannot choose but dig your own grave, and place your coffin in your eye.'

On Boxing Day, Hanzakos asked his sister, Constance, and her son and his family to drive down from Michigan to see him. They ate burgers at the LK Restaurant in town, inspected his grave, which he was proud of and which upset them, and then all crammed into Hanzakos's one-room apartment under a machine shop. The shoeshine handed out some jars of vegetables that he had canned and a few envelopes of cash. When his nephew tried to refuse the gift of his old shoe brushes, Hanzakos said, 'No, boy, I don't need anything any more,' took a step towards the kitchen table and fell to the floor. He was dead before the doctor arrived. The walls were decorated with ten calendars for 1960, the year that Hanzakos knew he would not see.

'The Man Who Died on Time', the story of Hanzakos's foreseen death, was published by *Life* magazine in early 1960. A few years later, it caught the eye of George Engel, a professor of psychiatry at the University of Rochester, in upstate New York, who collected such stories. Like Barker, Engel was gripped by cases of people who appeared to have dropped dead because of some surfeit of emotion or certainty about their fate. Barker spoke at Engel's hospital during his tour in May, and the psychiatrists swapped letters on his return to England. Some of their shared interests were uncanny. Engel often referred to the intemperate death of Dr John Hunter, whose ghost Barker searched for as a student at St George's Medical School. Engel was also greatly affected by the Aberfan disaster. When he showed a film about the tip slide during medical trials, in order to study the hormones of sadness, he often cried himself.

In the sixties, Engel collected 170 cases of sudden or eerie deaths, which he mostly found in press reports. He organised them into eight categories, including 'on loss of status or self-esteem', 'during acute grief' and 'after the danger is over' (a handful of people often die this way after earthquakes). Like Barker, Engel wanted to expand the frontiers of psychiatry and to pay more attention to the physiological impact of our emotions. In 1980 he wrote a landmark paper advocating a new 'biopsychosocial model' for medicine that would take into account not just the bodies of patients, but also their minds and the societies in which they lived. He was a man with time for the nocebo effect. And, like Barker, Engel was also compelled by forces that weren't entirely rational. On 11 July 1963, Engel's twin brother, Frank, who was also a distinguished doctor, died suddenly of a heart attack at the age of forty-nine. The two men had been exceptionally close: indistinguishable as boys; rivals and, more recently, collaborators as they pursued their medical careers. They called each other 'Oth', as in 'Other'.

234

After his brother died, Engel became convinced that his life was also running out. He saw death standing next to him. Then a magical notion took hold: that if he could survive the calendar year after his twin's passing, then he would live out a normal life. 'I was fully aware of the irrational nature of this idea, but nonetheless found it impossible to dispel,' Engel recalled. On the afternoon of 9 June 1964, just short of eleven months after his brother's death, and a few hours ahead of an awkward meeting which, like John Hunter, he was in no mood to attend, Engel suffered the heart attack that he had been waiting for. He was in his office in Rochester. It was not fear that he felt. 'My reaction to the attack was one of great relief. I not only escaped the unpleasant meeting, I no longer had to anticipate the heart attack; the other shoe had fallen, so to speak,' the psychiatrist wrote, in an extraordinary paper, in 1975. 'I felt serene and tranquil. The waiting was finally over.' I see and I know.

＊

Fairley once asked Barker to explain how he thought that people were scared to death. The psychiatrist said he thought that two mechanisms might be at work. 'I think suggestion is important,' Barker said. 'But on the other hand, I think that death at a certain time is predetermined.' His voice slowed and he chose his words with care. 'And therefore, shall we say, it is to a certain extent . . . fixed.' In other words, in cases of 'voodoo death' you might be able to frighten someone sufficiently to stop their heart. If they were a rat, you could crop their whiskers. But there is also the possibility that a person's death is simply ahead, waiting to happen. A warning doesn't bring it about. The future is already there. Some people catch a glimpse of their fate; most never will.

In Copthorne Hospital in Shrewsbury, the psychiatrist's theories closed in on him. Barker entered a state of mortal anticipation that he had studied and thought about for years. He once told Jane that he expected to die young. He did not seem afraid. Perhaps he thought he was being proved right. He verged on the great secret. He was part of the scheme of things.

During the eighteen months that the Premonitions Bureau was in operation, there had been plenty of signals that Barker was on an errant path. When he described the idea in the medical press, colleagues told him it was undignified and embarrassing. Even Littlejohn and the small-minded NHS officials who wanted to stop him publishing his book were probably, on some level, trying to protect his reputation. When a truly preventable tragedy in his life took place – a fire in an old hospital, full of alarms and hoses that nobody knew how to use – the bureau proved valueless. Ninety-seven per cent of its predictions did not come true. Its star percipients felt wronged and abused. A useful definition of a delusion is not that it is an inaccurate belief about the world; it is a belief that you refuse to change when you are confronted with proof that you are wrong. The hypothesis fails. The pleasure principle is countermanded by the reality principle. Our best hopes and most extravagant fears rarely materialise. Prediction errors fire through the brain, turning the tiger back into a shadow. Prophecy reduces to coincidence. Your heart rate slows. The experiment does not repeat. The pattern won't spread.

But what happens when the experiment happens to describe your life? And what if, in your case, the pattern does hold? Our lives are not tests available to be rerun. The Premonitions Bureau was right three per cent of the time and, by the end, Barker found that he was within the three per cent. Unusualness. Accuracy. Timing. There was an expectation that he would die. And there was an observable reality. I see and I know.

*

Barker was discharged from hospital in the middle of August. He was allowed to return to work, but advised to take it gently. He wasn't given an explanation for his headaches. Around a quarter of patients with subarachnoid haemorrhages are still misdiagnosed by doctors. They miss the signs. During his final week at Shelton, Barker toured the wards as usual. He was working on some new ideas about fear. Miss Middleton sent him a warning about the Duke and Duchess of Windsor. 'I have logged this prediction as usual!' he replied. On the Friday, Barker paid a home visit to a patient in Whitchurch, at the northern limit of his patch. The patient later remembered his 'extreme kindness, courtesy, and firm advice'. The following day, the psychiatrist waved off his friend Dr Enoch and his wife, who were going on a short holiday to Llandudno. It was the Barker family's last weekend in Yockleton. They were moving to the new house in a few days' time. The rest of the family were downstairs, on the Sunday morning, when they heard Barker's laboured breathing, sounding through the ceiling. He was on the bedroom floor. Jane went to him. A vessel had burst in his brain. He was conscious for a short time. 'Perhaps it is the apparent impossibility of it all that fascinated me,' he once wrote. In that moment, before the thunderclap, nothing was impossible at all. And then the future crashed in.

August 20, 1968

Dear Mr. Hencher,

I regret to have to inform you that
Dr. Barker died suddenly on Tuesday, August
20 last.

Yours sincerely,

Consultant's Secretary.

Mr. Alan P. Hencher,
27, Lodge Avenue,
DAGENHAM,
Essex.

Epilogue

John Barker died in hospital in Shrewsbury on 20 August 1968. His death, and the warning that preceded it, made the front page of the *Psychic News*. The morning before he died, Miss Middleton woke, again, with a choking sensation. She called out for help. After Barker's death, she sent most of her premonitions to the Central Premonitions Registry in New York. She died in 1999, looked after by Les Bacciarelli and surrounded by her cats. Peter Fairley became known as 'the face of space' for his work on British television, presenting the moon landings and chronicling NASA missions in the sixties and seventies. He died in 1998, at the age of sixty-seven. After Barker's death, Alan Hencher dropped out of contact with the bureau and moved to Suffolk. Jane Barker remarried happily and died in 2014, at the age of ninety.

The Premonitions Bureau continued to collect visions and forebodings from the British public into the seventies. Jennifer Preston catalogued them with care. She welcomed paranormal investigators and journalists to her house in Charlton, in south London, but longed for someone to take up the project seriously again. 'They are keen on hearing about premonitions or witches or poltergeists,' she told the *Evening Post* in 1973. 'But nobody seems to want to do anything constructive.' By the time the experiment ended,

Preston had a collection of more than three thousand premonitions, of which around 1200 had been checked, in filing cabinets in her house. No early warnings were ever issued.

1970s. J. Keith Cubbins's earlier memoir of working at Shelton, *The Waiting Room to Hell*, was also instructive. I am grateful to Sarah Davis and the staff of the Shropshire Archives for guiding me through the hospital records and for all your help during my pandemic-interrupted visits. Claire Hilton, historian in residence at the Royal College of Psychiatrists, was an invaluable source of knowledge, support and instinctive generosity throughout my research. Hilton's scholarship of the treatment of older mental patients, mental health reform and Barbara Robb's campaign group, AEGIS, underpins much of what I have written here. The archives of the Society of Psychical Research, at the University of Cambridge, held vital fragments of Barker's writings and correspondence that made me believe this book was possible.

Chuck Rapoport's experience and photographs of Aberfan were a constant source of inspiration, and I am honoured to reproduce two of them in the book. Jeremy Deedes, Bob Trevor, Magnus Linklater and David Johnson, in particular, helped to recreate the atmosphere and detail of 1960s Fleet Street and the vibrations of the *Evening Standard* newsroom. I hope it smells right. I owe many debts to individual researchers who provided information or analysis that I could find nowhere else. They include Owen Davies and his book, *A Supernatural War*, Lucy Noakes's *Dying for the Nation: Death, Grief and Bereavement in Second World War Britain*, Elizabeth Rottenberg's *For the Love of Psychoanalysis: the Play of Chance in Freud and Derrida*, and Asif Siddiqi's

Acknowledgements

I will always be grateful to the children of John Barker – Nigel, Josephine, Julian and Simon – for trusting me to write your father's story. You shared photographs, letters, report cards, old audio reels and precious childhood memories to allow me to put together this account. The mistakes and missteps are mine. I can only hope to have made an honest portrait. Thank you.

Duncan and Alastair, the sons of Peter Fairley, also went out of their way to dig out photographs and books and any remaining traces of the Premonitions Bureau. Likewise Jonathan and Arabella Preston, children of Jennifer Preston, who guarded and maintained the bureau for years after the loss of the project's impetus. I am grateful to the relatives of Alan Hencher, who confirmed important details about his upbringing, and I would never have been able to do any kind of justice (if indeed I have) to the life and mind of Miss Middleton without the help of Derek Sunners and Miss Middleton's long-standing friend, pupil and neighbour, Christine Williams.

Surviving staff of Shelton Hospital, including Rosie Morris, David Enoch, Robert Quinlin and Harry Sheehan, provided me with priceless, ephemeral memories of daily life and ward rounds in the hospital in the 1960s and early painstaking analysis of Vladimir Komarov's ill-fated flight on Soyuz-1. Benjamin Wieland, an editor at CH Media, in Basel, went above and beyond in helping me to understand the circumstances of the Globe Air crash in Cyprus in 1967. Larry Tye shared his sense of Bobby Kennedy's understanding of fate.

Luana Colloca, Fabrizio Benedetti, Ted Kaptchuk, Giulio Ongaro and Shelly Adler helped me to understand the nocebo effect. Peter Kennedy told me about the background of his father, Walter's, research and naming of the phenomenon. I drew heavily on the reporting of my friend and colleague, Rachel Aviv, to describe resignation syndrome in Sweden and on the insights of Karl Sallin into the condition. Philip Corlett patiently entertained my questions about prediction, perception and delusions. Martin Samuels introduced me to the work of George Engel and explained his own pathbreaking neurological research on the brain and its relations with our other main organs, including the heart. Everyone was open and curious and took Barker's preoccupations seriously.

When you attempt to write about time and our place in it, you work in the shadow of wonderful writers and I have been grateful, when not daunted, by the brilliance of Annie Dillard, Carlo Rovelli, J. B. Priestley, Charlotte Beradt, John Berger, Oliver Sacks, Marina Warner, Andy Clark, W. G. Sebald, Arthur Koestler, Philippe Ariès, Janet Malcolm and John Gray, among others, when thinking about the themes of this book. Teresa Gleadowe, at CAST, in Helston, and

Rohan Silva, at Second Home, in London, gave me places to work. Samara Clarke supplied time and unfailing, loving support to my family during the pandemic.

I have been fortunate to work with outstanding magazine editors – Alice Fishburn, Alan Burdick, Chris Cox, Jonathan Shainin and David Wolf – all of whom have known about this story, in various iterations, and encouraged me to pursue it. I am grateful to my colleagues at the *New Yorker*, David Remnick, Dorothy Wickenden, Daniel Zalewski and Willing Davidson for commissioning and publishing the magazine article, 'The Premonitions Bureau', in March 2019, and then giving me time off to write the book. Zach Helfand checked the original story and made sure I was on solid ground. The Invisible Bureau has only ever been a WhatsApp away. Willing, Gideon Lewis Kraus, Ben Power, Sue Williams, Will Carr and AC Farstad read early drafts of the book and have been constant sources of advice and friendship. Jonathan Heaf, Ed Caesar, Mark Richards, Rose Garnett, Rebecca Servadio, Tom Basden and Emily Stokes steered me right at critical moments. Peter Straus rang me out of the blue in 2006 and told me to write a book. I am sorry it has taken so long.

Will Heyward, at Penguin Press, and Alex Bowler, at Faber, understood immediately the kind of book that I wanted to write and have challenged me, in the best way, to realise it. Natalie Coleman and Anne Owen have made sure that it has actually happened. Thank you. At Aitken Alexander, Lesley Thorne has always had my back, and it is hard to convey my debt to Emma Paterson, who has guided my work and my writing almost (but not quite) without me realising it, since we met in 2014. You are wise, kind and lethal. I am lucky to know you. The love of my parents, Bill and Stephanie, and my sister, Sarah, supports me every day, and more than they realise. I acknowledge my children: Aggie, Tess, John and Arthur. I love you very much. I am grateful to whatever magical force, occult or real, that has enabled me to share my life with my wife, Polly. This book is for you.

Illustrations